xtbw 10-30-87 (11)   ob   1-19-88

ACS SYMPOSIUM SERIES 357

Experimental Organometallic Chemistry

A Practicum in Synthesis and Characterization

Andrea L. Wayda, EDITOR
AT&T Bell Laboratories

Marcetta Y. Darensbourg, EDITOR
Texas A&M University

Developed from a symposium sponsored
by the Division of Inorganic Chemistry
at the 190th Meeting
of the American Chemical Society,
Chicago, Illinois,
September 8–13, 1985

American Chemical Society, Washington, DC 1987

Library of Congress Cataloging-in-Publication Data

Experimental organometallic chemistry: a practicum in
synthesis and characterization / Andrea L. Wayda,
editor, Marcetta Y. Darensbourg, editor; developed
from a symposium sponsored by the Division of
Inorganic Chemistry at the 190th meeting of the
American Chemical Society, Chicago, Illinois,
September 8-13, 1985.

　　p.　　cm.—(ACS symposium series, ISSN
0097-6156; 357)

　　Includes bibliographies and indexes.

　　ISBN 0-8412-1438-7

　　1. Organometallic chemistry—Congresses.

　　I. Wayda, Andrea L. (Andrea Lynn), 1954-
II. Darensbourg, Marcetta Y., 1942-　　. III. American
Chemical Society. Division of Inorganic Chemistry.
IV. American Chemical Society. Meeting (190th: 1985:
Chicago, Ill.) V. Series.

QD410.E97　　1987
547′.05—dc19　　　　　　　　　　87-27353
　　　　　　　　　　　　　　　　　　　CIP

ACS Symposium Series

M. Joan Comstock, *Series Editor*

Foreword

The ACS SYMPOSIUM SERIES was founded in 1974 to provide a medium for publishing symposia quickly in book form. The format of the Series parallels that of the continuing ADVANCES IN CHEMISTRY SERIES except that, in order to save time, the papers are not typeset but are reproduced as they are submitted by the authors in camera-ready form. Papers are reviewed under the supervision of the Editors with the assistance of the Series Advisory Board and are selected to maintain the integrity of the symposia; however, verbatim reproductions of previously published papers are not accepted. Both reviews and reports of research are acceptable, because symposia may embrace both types of presentation.

Contents

Preface

THE SYNERGISTIC RELATIONSHIP BETWEEN technique development and new materials development is more clearly evident in inorganic and organometallic chemistry than in any other area of chemistry. In these fields, the accessibility of new compounds derived from highly reactive starting materials as well as the ability to manipulate molecules unstable in ambient conditions has further encouraged chemists to develop and refine spectroscopic probes and characterization techniques. These techniques have in turn yielded details of structure and bonding that have led the synthesis chemist to explore an ever-expanding array of atom–atom connections, leading to the synthesis of novel materials and compounds. As important as the above process is the insight gained into reactivity trends and reaction mechanisms whenever advances are made in identifying and monitoring reaction intermediates and products. With continued synthetic refinements and detection and characterization of more and more "transient" species, or stable relatives of transient species, understanding of individual steps in a reaction pathway and ultimate control of product distribution become more reachable goals. As this sequence of events has been reiterated, the problems and the sophistication of our science have increased, and the applicability of our results has extended far beyond the classical boundaries of the inorganic division of chemistry. So much activity, so many new research methods, and so many possibilities have converged to create the need for organized, focused educational efforts such as tutorials and symposia.

The idea for the organometallic methods tutorial from which this book was developed grew from a discussion about the demanding synthetic and characterization skills necessary to ensure the isolation and structural elucidation of anionic organometallics. Encouraged by Du Shriver, then chair of the Division of Inorganic Chemistry, we organized a tutorial in the form of a practicum that complemented the symposium on anionic organometallics held at the ACS National Meeting in Chicago. The enthusiastic comments of the audience and the overflow attendance at the tutorial led to the decision to publish the proceedings (supplemented with additional topics) as a volume in the ACS Symposium Series.

The book is structured around the two major themes of the tutorial: synthetic methodology and characterization techniques. Chapters correspond to papers presented at the tutorial and several other subjects (invited contributions that considerably expand the scope of the original choice of topics selected for the tutorial). The ancillary and intended functions of the book are as follows:

- to gather, in greater detail than journal publications permit, specialized techniques for difficult syntheses, as well as handling and sampling techniques that have worked well for some of our leading experimentalists (with examples illustrating their specific problems) and

- to provide a working knowledge of selected characterization techniques that are currently important to the further development of inorganic and organometallic chemistry. Both familiar and less well-known techniques are presented to permit a critical assessment of their applicability and reliability.

The major chapters form the backbone of the book. They are buttressed by short contributions entitled Applications, which provide specific detail or document a useful twist or adjunct to the described topic. This organizational scheme closely parallels other works on the topic (*1-6*). However, what distinguishes this volume from others in the genre, and in our opinion makes it unique, is the absence of a single authoritative voice in favor of a consortium of contributors and views. Authors throughout the organometallic community were solicited by direct mail; contributions were sought not only from the United States but also from abroad in an effort to represent the broadest possible spectrum of organometallic techniques and methods. Our aim has been to avoid the covert scientific relativism (*7*) that occasionally creeps into our thinking when we have grown accustomed to a particular way of "doing chemistry" in a specific scientific and cultural milieu. Unfortunately, because the response was greatest from U.S. chemists and because we exert our own idiosyncratic influence on topic choice, a certain parochiality of view is likely to remain. However, this approach has produced a book of techniques and methods with a strikingly different slant than other books of this type.

This claim is best illustrated by the synthetic section of the book. Organometallic chemists who have been trained in classical traditions with slightly different emphases on how to best exclude air and water from sensitive reaction chemistry (favoring Schlenk, vacuum line, glovebox, or combined methodologies) present their contrasting individual solutions to common synthetic problems. The effect is to accentuate the firm ground on which these techniques are based (yet at the same time preserve the individuality of the different approaches) and to confirm the variety of viable methodology available to the open-minded and flexible synthetic organometallic chemist.

A group of contributions devoted to high-technology synthetic strategies bridges the sections devoted to synthesis and characterization.

High-technology synthetic strategies are on the interface of materials science and physical chemistry; they are a topic of growing interest to organometallic chemists interested in nontraditional applications. The book then reviews state-of-the-art characterization techniques most likely to be useful to practicing organometallic chemists: FTIR methods, metal and high-pressure NMR spectroscopy, and X-ray diffraction analysis. The book concludes with a chapter on the rapidly developing area of photoelectron spectroscopy. This contribution highlights the importance of this relatively new technique in constructing modern structure–property relationships in organometallic chemistry.

If we have been successful in our aims, the reader should be able to scan the following chapters and choose the methods or characterization techniques that best suit the chemistry at hand. She or he should at the same time acquire an appreciation for less familiar methodology (or at least be made aware of their existence and potential use in the future). Finally, we hope that these accounts, written from the viewpoint of organometallic chemists, spectroscopists, and theoreticians who are enthusiastic innovators in the topics described, will provide the interested reader with the sense of art that is implicit in synthetic organometallic chemistry but that of necessity is sadly lacking in the concise detail of the experimental sections of our papers (8,9).

A Note About Safety

The contributors to this volume were asked to indicate hazardous procedures and safety precautions to be followed when appropriate. To the best of their knowledge and experience, and to ours, this has been done conscientiously. Nevertheless, every set of conditions generated in a particular laboratory cannot be predicted. This book assumes a level of expertise of the practicing professional chemist. No new procedure should be attempted without a thorough investigation of health hazards and disposal protocol associated with all starting reagents and possible products. No new apparatus should be assembled and used without a thorough safety inspection. We urge you to exercise the caution your profession demands.

Acknowledgments

We thank the individuals who have been instrumental in the preparation of this volume: Du Shriver for his continuing interest in the project and for contributing Chapter 1 on the historical development of techniques in organometallic chemistry; the members of the Division of Inorganic Chemistry for their unflagging support before and during the realization of this project; Robin Giroux for providing helpful education in the logistics of editorial responsibilities and other practical matters; and most important, the scientists who so willingly, openly, and enthusiastically shared their expertie and "tricks of the trade" with us—and with you. Finally, A. L. W. acknowledges an enormous debt to her stalwart philosophical collaborators H. D. Keith, M. L. Kaplan, and T. M. Wolf of AT&T Bell Laboratories; J. L. Dye of Michigan State University; and in particular, A. A. Rosenfeld of the University of Chicago for providing much-needed encouragement and moral support during the production of this book.

References

1. Shriver, D. F.; Drezdon, M. A. *The Manipulation of Air-Sensitive Compounds,* 2nd ed.; Wiley: New York, 1986.
2. Gysling, H. G.; Thunberg, A. L. In *Physical Methods of Chemistry*; Rossiter, B. W., Ed.; Wiley: New York, 1986; pp 373–487; Vol. 1, *Components of Scientific Instruments and Applications of Computers to Chemistry.*
3. Yamamoto, A. *Organotransition Metal Chemistry: Fundamental Concepts and Applications*; Wiley: New York, 1986; Chapter 5.
4. *Organometallic Syntheses*; Eisch, J. J.; King, R. B., Eds.; Academic: New York, 1965; Vol. 1, King, R. B., 1965; Vol. 2, Eisch, J. J., 1982.
5. Brown, H. C.; Kramer, G. W.; Levy, A. B.; Midland, M. M. *Organic Syntheses Via Boranes*; Wiley-Interscience: 1973; Chapter 9, pp 191-261.
6. Herzog, S.; Dehnert, J.; Luhder, K. In *Technique of Inorganic Chemistry*; Jonassen, H. B.; Weissberger, A., Eds.; Wiley-Interscience: New York, 1968, p 119.
7. *Cultural relativism* is defined as the interpretation of experience and knowledge in the light of a specific personal and cultural perspective (Herskovits, Melville. *Cultural Relativism: Perspectives in Cultural Pluralism*; Random House: New York, 1972; p 15). *Scientific relativism* may be defined analogously.
8. This connection is very clearly drawn by Jacob W. Getzels and Mihaly Czikszentmihalyi in *The Creative Vision: A Longitudinal Study of Problem Finding in Art*; Wiley-Interscience: New York, 1976. A provocative collection of recent essays on the topic may be found in "Art and Science"; *Daedalus,* **1986,** *115(3).*
9. For a specific discussion of the role journals play in the communication of scientific results, see *Little Science, Big Science*; Columbia University Press: New York, 1963 by Derek J. de Solla Price. A treatment of the issue in the broader context of science as a subset of the sociology of knowledge may be found in Robert K. Merton's *The Sociology of Science*; University of Chicago Press: Chicago, 1973.

Andrea L. Wayda
AT&T Bell Laboratories
Murray Hill, NJ 07974

September 29, 1987

Marcetta Y. Darensbourg
Department of Chemistry
Texas A&M University
College Station, TX 77843

Chapter 1

Development of Techniques in Organometallic Chemistry

D. F. Shriver

Department of Chemistry, Northwestern University, Evanston, IL 60201

As in all physical sciences, advances in organometallic chemistry have been strongly coupled to advances in techniques. The field is flourishing in large part because of the development of versatile, user-friendly NMR and IR spectrometers and X-ray diffractometers. Similarly, steady improvements in the techniques for synthesizing and handling organometallics have greatly increased the efficiency of our research.

Many organometallics are air sensitive, so the development of techniques which permit synthesis and characterization in a vacuum or inert atmosphere are central to the field. Even though the manipulation of air-sensitive compounds has been successfully practiced for a long time, the techniques have undergone constant improvement. For convenience these may be classified as vacuum line, bench-top inert atmosphere, and glove box techniques but as shown in this book combinations of them are common.

One of the most elegant and rigorous methods for handling volatile air-sensitive materials is the chemical vacuum line, which was developed in the early 1900's by Alfred Stock for the synthesis of nonmetal hydrides, including those of boron and silicon. When Stock began these investigations around 1902 at Berlin, the rotary oil-sealed mechanical vacuum pump had been recently developed, but the assembly of the rest of the apparatus was not as simple. For example, the complex series of traps and valves, which Stock eventually devised, Figure 1, had to be constructed on the spot from soft glass, material which is highly subject to breakage by thermal shock.

Stock's designs for chemical vacuum line apparatus were widely disseminated. A leading American inorganic chemist, Professor L. M. Dennis at Cornell University became familiar with Stock's apparatus in the course of trips to Europe and he utilized chemical vacuum lines to investigate the hydrides of germanium in the late 1920's(1). About this same time chemical vacuum line techniques were

0097–6156/87/0357–0001$06.00/0

introduced to the U.S.A. for the investigation of boron hydrides by
Anton Burg, who worked   at the University of Chicago, first as a
student with Schlesinger and later a staff member. Burg acquired
these techniques by carefully reading Stock's papers and adding
innovations of his own(1).   Several other former Schlesinger
students, such as H. C. Brown, Riley Schaffer, and Grant Urry, have
also made major contributions to the development of vacuum-line and
inert-atmosphere apparatus.

In addition to a source of vacuum, a chemical vacuum line may
include a series of U-traps to separate compounds of different
volatility, reaction vessels of various types, and provision for the
storage of volatile solvents and gases. Vacuum systems of this type
have been extensively used to prepare and characterize low molecular
weight organometallics. Some of the major problems in vacuum line
work have been the erosion of stopcock grease by nonpolar solvents
and the mechanical inflexibility of the apparatus.

In recent years these problems have been reduced by the use of
O-ring joints, valves with Teflon stems and glass bodies, and
flexible stainless steel tubing. The chapter by Andrea Wayda in this
volume presents an excellent example of how these components can be
utilized in an apparatus with provision for both high vacuum and
inert-atmosphere manipulations.

In contrast with high-vacuum lines, inert-atmosphere apparatus
is more often used to handle liquids than gases or vapors. As with
the chemical vacuum line, inert-atmosphere bench-top techniques were
originally developed in Europe, but there is no single originator for
this diverse set of apparatus and techniques. The reaction tube with
a side arm for the   introduction of inert gas, Figure 2a, was
described by Walter Schlenk in 1913, and is still in use today but it
has been transformed to a more convenient form by modern O-ring
joints and Teflon-glass valves, Figure 2b.

In the late 1960's Professor James Burlich and I, as consultants
to Ace Glass Co. and Kontes Glass Co. respectively, assisted with the
commercial introduction of complete lines of inert-atmosphere glassware. From these companies and, more recently, Aldrich Chemical
Co., it is possible to order a inert-atmosphere setup all the way
from an inert gas purifier to filters and Schlenk flasks.

A third major technology, the inert-atmosphere glove-box was
developed in its modern form during World War 2 in connection with
fission weapons programs. Glove boxes are especially suited for
handling solids, but they also are widely used in the U.S.A. for
solution chemistry. Excellent inert-atmosphere glove boxes have been
commercially available for many years.

One of the reasons that inert-atmosphere techniques continue to
evolve is the availability of new materials. The introduction of
borosilicate glass in the 1920's led to great improvements in vacuum
lines and all types of laboratory glassware. Modern O-ring joints
would be virtually useless if it were not for synthetic rubber
O-rings, that are resistant to a reasonable range of solvents.
Similarly, butyl rubber, which is durable and has a relatively low
permeability to air, is used as a glove material for inert-atmosphere
boxes. Teflon has made the construction of greaseless valves possible, and in many applications these have replaced greased stopcocks
and Stock's expensive and cumbersome mercury float valves. Some new

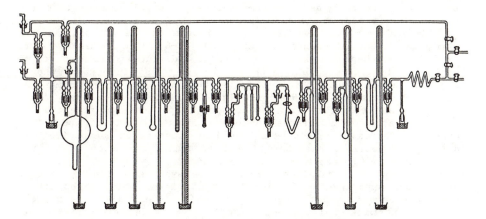

Figure 1. An early, <u>ca</u>. 1913, design for Stock's vacuum line. The items containing a pair of darkly shaded floats are mercury float valves. The mercury reservoir below each of these has been omitted. The sources of vacuum and inert atmosphere also have been omitted.
(Reproduced with permission from Ref. 7. Copyright 1933 Cornell University Press.)

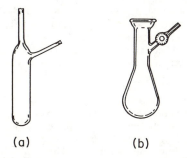

(a)   (b)

Figure 2. (a) The original design for a Schlenk tube. (b) A modern Schlenk tube with grease-free joint and valve and pear-shaped chamber to facilitate stirring and solvent removal under vacuum.

materials which should lead to useful new equipment designs are ceramics that can be shaped by machining, plastics with high chemical resistance and high service temperatures, and elastomers that are resistant to a wide range of solvents.

Several reviews describe the equipment and techniques for the manipulation of air-sensitive compounds.(2-6) The present volume differs from these by the presentation of individualistic accounts of equipment and techniques. Both the beginner and the expert should find many useful ideas here. Bear in mind, however, that the manipulation of air-sensitive compounds is less flexible than open beaker chemistry and therefore it is necessary to pay careful attention to the integration of apparatus so that a compatible set of operations can be performed. This can be achieved by planning each experiment in detail. Sometimes, sketches of the apparatus at each step in a synthesis serve to catch potential problems.

The expert in handling air-sensitive systems should be able to design new apparatus, and construct small items which might otherwise take days to run through a glass shop. Versatility is a great virtue in this field. For example, some chemists consider it impossible to start serious research on air-sensitive compounds before a costly dry box system has been acquired. This view contrasts with that of a leading European laboratory, which I visited recently. A dry box that hadn't been used in several years sat in a corner, even though very air-sensitive compounds were being synthesized and characterized in that laboratory. To emphasize the quality of their inert-atmosphere techniques, my host displayed a sealed vial containing a beautiful crystalline compound which his co-worker had prepared. He then cracked the vial open to demonstrate that the compound bursts into flames when exposed to air. Clearly, the well designed inert-atmosphere glassware, and good laboratory techniques practiced in that laboratory are perfectly satisfactory and probably more efficient than the use of glove-boxes.

The splendid collection of techniques presented in this volume should provide a stimulus to further innovations.

Literature Cited

1. I appreciate correspondence with Professors E. G. Rochow and A. B. Burg who provided information about developments at Cornell University and the University of Chicago.
2. Gysling, H. G.; Thunberg, A. L. In Physical Methods of Chemistry, Vol 1: Components of Scientific Instruments and Applications of Computers to Chemistry, Rossiter, B. W., Ed.; John Wiley Inc.: New York, 1986; p 373.
3. Shriver, D. F.; Drezdzon, M. A. The Manipulation of Air-Sensitive Compounds, 2nd edn, John Wiley Inc.: New York, 1986.
4. Eisch, J. J.; King, R. B., Eds.; Organometallic Syntheses, Academic: New York, Vol. 1 by R. B. King, 1965; Vol. 2 by J.J. Eisch, 1982.
5. Brown, H. C.; Kramer, G. W.; Levy, A. B.; Midland, M. M. Organic Synthesis via Boranes, Wiley-Interscience Inc.: 1973, Chapter 9; p. 119.

6. Herzog, S.; Dehnert, J.; Luhder, K.   In In Technique of Inorganic Chemistry, Jonassen, H. B.; Weissberger, A., Eds.; Wiley-Interscience Inc.: New York, 1968; p. 119.
7. Stock, A.   The Hydrides of Boron and Silicon; Cornell University Press: Ithaca, NY, 1933.

RECEIVED September 1, 1987

Chapter 2

Cannula Techniques for the Manipulation of Air-Sensitive Materials

John P. McNally [1], Voon S. Leong [1], and N. John Cooper [2,3]

[1]Chemistry Department, Harvard University, Cambridge, MA 02138
[2]Chemistry Department, University of Pittsburgh, Pittsburgh, PA 15260

Techniques based on solution transfer through stainless
steel or Teflon cannulae permit rapid, convenient manipulation of air sensitive materials under an inert
atmosphere. The basic equipment required for cannula
manipulations is described, and their application to
routine laboratory procedures such as recrystallization, sublimation, chromatography, and the preparation
of spectroscopic and analytical samples is discussed.

The development of modern organometallic chemistry has depended on
technical advances which have made it feasible and convenient to
routinely manipulate air sensitive and water sensitive materials in
the synthetic laboratory.[1] Key components of these advances include
cheap, reliable pumping systems capable of providing medium or high
vacuums, and the routine availability of high quality nitrogen and
argon. These components may be combined in many ways to provide
inert atmosphere working conditions, and some of those approaches,
such as the use of high vacuum lines derived from Stock vacuum lines
and inert atmosphere glove boxes, will be discussed in other sections of this book.
   Unfortunately many handling techniques, while they provide
excellent inert atmosphere conditions, either require fairly complex
and expensive capital equipment, or divorce the synthetic chemist
significantly from the convenience and flexibility of traditional
flask and funnel synthetic operations. The closest approach to traditional handling techniques is provided by systems based on Schlenk
tubes and commercially available variations such as the Ace No Air
apparatus,[2] which depend on the rigid assembly of combinations of
Schlenk tubes and fritted glass filtration apparati connected by
standard taper ground glass joints. An early variation on these
techniques was the use of syringes to inject reagents and solvents
into reaction vessels through reusable rubber serum caps,[3] and more

[3]Address correspondence to this author.

0097–6156/87/0357–0006$06.00/0

recently researchers from many laboratories have realized that light, stainless steel cannulae allow air sensitive handling techniques which have most of the physical and manipulative flexibility of traditional synthetic equipment.

Cannula techniques have several advantages over other approaches to the manipulation of very sensitive materials:

1. They are quicker than other manipulative systems, since they do not require the careful setup characteristic of systems based on Schlenk tubes and fritted glass filters, and the chemist is not impeded by the heavy rubber gloves characteristic of glove box operations.
2. The techniques are more flexible and require less detailed planning of the intended experiment, allowing the experimenter to change the purification approach relatively easily.
3. A bench can be equipped with the apparatus necessary for cannula techniques for a relatively modest sum.
4. The techniques can be learnt quickly by inexperienced personnel.
5. Once the basic equipment is available, it is straightforward to adapt the techniques to many conventional laboratory manipulations.

Cannula techniques are complementary to other techniques for handling air sensitive materials. In particular, while they may be used to prepare NMR samples of air sensitive materials and to manipulate solids, it is normally more convenient to carry out both these operations in an inert atmosphere glove box. In our experience cannula techniques are at least as good as any other technique for the exclusion of oxygen from a reaction mixture (in fact we find them to be superior to inert atmosphere glove boxes), but glove boxes and high vacuum manipulation lines provide better protection from adventitious water.

The objective of the present chapter is to describe the combination of cannula/Schlenk tube techniques which are currently used in our own laboratories for the manipulation of air sensitive materials. These techniques are not, of course, unique to our laboratories - similar techniques have been developed in many other groups, and many of the techniques to be described were initially developed elsewhere, or are based on conversations with colleagues.

Basic Equipment

Vacuum Line. The core of cannula/Schlenk techniques is the vacuum line shown in Figure 1 and Figure 2. This design has evolved over a number of years and it has a number of advantages over other commonly used designs:

1. It is cheap to build. Standard 4 mm, 3-way oblique bore stopcocks can be used for the main stopcocks, although precision ground hollow plug stopcocks offer better performance. The less precise grinding of solid plug stopcocks can be compensated for by finishing the grind in situ using 400-1000 grit silicon carbide grain. This has the disadvantage that the interchangeability of the plugs is lost.

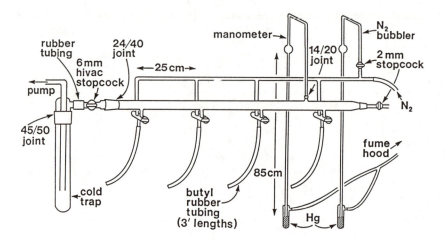

Figure 1.   Vacuum line for cannula manipulations.

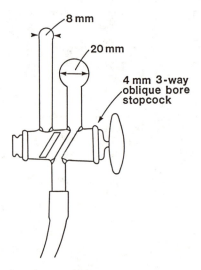

Figure 2.   Cross section of vacuum line.

2. The line is easily taken apart for cleaning. The main vacuum manifold is readily cleaned with a burrette brush through the 24/40 joint.
3. Rubber vacuum tubing is used to connect the trap section to the rotary pump and to the vacuum manifold. The cold trap is typically emptied many times a day, and the flexibility of this connection reduces breakage.
4. The trap is connected to the vacuum line in the reverse of the normal sense – in most vacuum systems volatiles enter the trap through the central tube; when using liquid nitrogen coolant and pumping off large volumes of solvent, as is typical in cannula manipulations, this results in frequent blockages and it is more convenient to connect the cold trap as shown. A second trap may be added to prolong the life of the pump oil.
5. The use of relatively narrow bore tubing for the nitrogen supply system results in a small dead volume in the system, and minimizes contamination when nitrogen tanks are changed, etc.
6. The system uses a simple rotary pump and no diffusion pump is required. Any dual stage pump with a free air displacement of 20 liters per minute or more should be satisfactory, although larger free air displacements are advantageous in terms of routine working speed. We have found Leybold Heraeus model D2A direct drive pumps with free air displacements of 62.5 liters per minute to be particularly satisfactory, and very quiet in routine operations.
7. Any convenient weight of vacuum tubing can be used to connect the stopcocks to the Schlenk tubes, but we have found 1/4" bore 1/8" wall butyl rubber tubing (Fisher) to have significant advantages over natural latex tubing. In addition to reduced oxygen permeability, the butyl rubber tubing absorbs less water, and is more resistant to oxygen degradation, and has a longer lifetime.

Schlenk Tubes. Routine manipulations are carried out in Schlenk reaction vessels, and Figure 3 shows the dimensions of the flasks most commonly used in our laboratories. Most of the vessels use a male 24/40 joint; the advantage over a female joint is that it reduces leaching of grease into the reaction vessel. We do, however, find female joints to be convenient for the smaller scale 50 mL and 10 mL reaction vessels. We also find 800 mL vessels to be convenient solvent reservoirs and we have used vessels up to 2.5 L in size for large scale preparations.

Cannulae. Transfer cannulae (double tip needles) are hypodermic grade stainless steel tubes which may be purchased from Aldrich (with noncoring tips) or from Popper and Sons (unfinished). We find 3 feet lengths to be most convenient, and the most useful sizes are 16 gauge, 18 gauge, 20 gauge. We also find 22 gauge to be convenient for some purposes, although it is more prone to blockage than the wider bores.
   Most recently we have started to make extensive use of Teflon tubing for the transfer of liquids; this has the advantage that the wider bore Teflon tubing is more flexible and offers less resistance to solution flow. The tubing (Daiger Scientific) is 1/8" id and

1/32" wall. The only disadvantages of Teflon tubing are that the
serum caps must be drilled with a cork borer, and that accidental
"kinking" reduces the lifetime of the transfer tubing.

Septa. Two types of serum caps (septa) have been commonly used for
cannula manipulations; the older style have a smooth wall, but more
recently Suba Seal septa with a ridged wall which provides better
closure have become commercially available in this country. Suba
Seals are available from Aldrich and from Strem, and in addition to
providing a better seal they are made of harder rubber, have better
solvent resistance, and can be penetrated more frequently before
deteriorating.

Filter Stick. A typical filter stick for solution filtration is
shown in Figure 4. A section of teflon tubing (2 - 3 ft) is pushed
over the Luer fitting of an observation port (Popper & Sons), and
conventional filter paper is then attached to the end of the observation port as shown. Half inch teflon sealing tape (Aldrich - this
may also be available from a local plumbers' supplier as pipe dope)
is first wrapped around the base of the observation port, filter
paper is then wrapped around the outside of the teflon tape, and one
or two loops of nichrome wire (20 gauge, 32 thou) are used to secure
the filter paper in place. The teflon tape provides a bed into
which the wire can bite and seal the system. Any filter paper may
be used, but we have found water resistant fine filter papers such
as Whatman's #50 to be most satisfactory, except for gummy solids
which may require a faster paper.

Grease. Conventional silicone high vacuum grease is the most convenient lubricant for the vacuum line stopcocks, but we find Krytox
240 AC, a fluorinated grease developed for satellite and missile
applications, to have significant advantages over silicone grease
for stopcocks on the Schlenk tubes. The wetting properties of
Krytox are not quite as good as those of silicone high vacuum
greases, and Krytox tends to streak at low temperatures ($< -40°C$),
but it is much more resistant to leaching by non-aqueous solvents
and its use essentially eliminates contamination of samples by
grease.

Principles of Operation

Pump and Purge. A key objective of cannula techniques is to achieve
very low partial pressures of oxygen without the use of time consuming high vacuum techniques, and this is achieved by repeated pump
and purge cycles. Even a heavily used rotary pump will give an
ultimate vacuum of 1×10^{-2} mm Hg, and will rapidly reduce the
pressure in a small reaction vessel to 1 mm Hg. In a typical
example a 224 mL reaction vessel contains approximately 2 mmol of
O_2, and evacuation to 1 mm Hg reduces the quantity of O_2 to 2×10^{-3}
mmol. If the vessel is refilled and the cycle repeated the quantity
of O_2 reduces to 2×10^{-6} mmol, and it is clear that relatively poor
vacuums are sufficient to provide very low effective partial
pressures of O_2 after 2 or 3 pump/purge cycles.

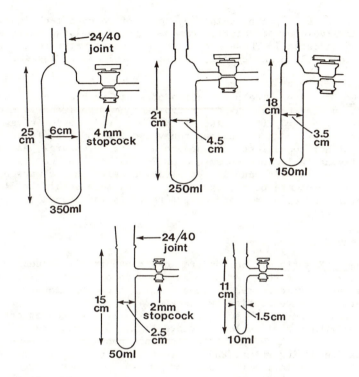

Figure 3.   The most useful Schlenk tube sizes for routine synthetic work.

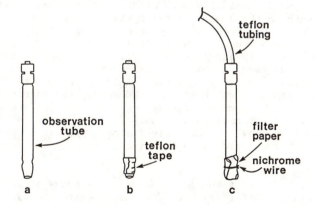

Figure 4.   Conversion of an observation port into a filter stick. The luggs can be sawed off the Luer fitting on the observation tube to make a more streamlined apparatus. A second loop of nichrome wire may be convenient to prevent leakage.

Water Reduction. The principle weakness of cannula techniques is that small volumes of solution can be exposed to large surface areas of glass and steel. These are major sources of water contamination, and we routinely oven dry all transfer cannulae and glassware before use.

Positive Pressure. The penetration of septa by transfer cannulae results in a system which is far from vacuum tight, and to minimize contamination by O_2 it is essential to work under a positive pressure of an inert gas unless the vessel is under a virgin septum. It is neither convenient nor safe to use high pressures, and we typically use a low over-pressure of 2-4 psi. When two vessels are to be connected by a cannula the first vessel should always be under a positive pressure of nitrogen, and the cannula should be flushed with nitrogen for a few seconds before it is introduced into the second vessel.

Basic Manipulations

Setting up Reactions. Solids are most conveniently loaded in a glove box, or, if this is not available, in a reusable glove bag. Solids may be transferred between Schlenk tubes by means of an adapter made by connecting two female 24/40 joints by a short tube with a 120° bend in it, but this is typically less convenient than transfer within a glove box or glove bag. Solvents are collected from a still in which they have been continuously refluxed over an appropriate drying agent, and are pumped and purged with nitrogen as described above to remove traces of oxygen. The dry, oxygen-free solvent may then be transferred onto the solid under positive pressure of inert gas (as in Figure 5, with a cannula in place of a filter stick). This basic liquid transfer is used many times, and the transfer is typically accomplished using a small over-pressure of nitrogen on the source vessel while the pressure in the receiving vessel is allowed to equalize with atmospheric by means of an exit needle in the serum cap. We typically use 2" 18 gauge needles as exit needles, although longer needles may be used if there is some concern about back diffusion of oxygen into the reaction vessel. Reagents may be added in a similar fashion, or may be added via syringe. Reagents which are not themselves air sensitive may often be conveniently spooned directly into the reaction vessel against a countercurrent of nitrogen. A surprisingly gentle countercurrent is sufficient to prevent back diffusion of oxygen into the reaction vessel over the period of a few minutes required to complete addition of the reagent.

Filtration. One of the most convenient aspects of the cannula techniques is that filtration does not require the use of fritted glass equipment. Small scale filtrations are simply carried out by transferring the liquid to be filtered from one Schlenk tube to another through a filter stick as shown in Figure 5. When the filtration is complete the filter paper is discarded, and the cannula washed with acetone to prepare it for the next filtration. The advantages of filter stick filtration over filtration through fritted glass apparatus maybe summarized as follows:

1. Repeated filtrations are much faster, since it takes a matter of minutes to discard the old filter paper, clean the filter stick, and wire on a new filter paper.
2. Small quantities of solids are much more conveniently collected by filter stick filtration, since mechanical losses are small with the small surface area of the filter stick. This can be a major advantage when working with some of the more expensive metals, since it becomes convenient to carry out exploratory synthesis on scales as small as 50 mg.
3. The flexibility of the apparatus minimizes breakage.
4. Filter stick filtration is particularly convenient for collecting crystals from cold or hot solutions since it can be carried out at any convenient tem perature without the use of special apparatus.
5. Filtration requires minimal equipment.
6. Elimination of the need to clean filter frits can save a great deal of experimental time.

The only disadvantage of filter stick filtration is that the small surface area of the filter may result in slow filtration with very fine percipitates or gelatinous precipitates. This typically limits filter stick filtration to reactions on a scale of 2 g or less starting material.

Larger scale filtrations are conveniently carried out using fritted glass filter apparatus, and a suitable design for Schlenk tube work is shown in Figure 6. The key features are the use of 24/40 joints to ensure compatibility with the Schlenk ware, and the use of a greaseless high vacuum stopcock. This minimizes contamination of the products by vacuum grease, and we have found Youngs taps (supplied by Brunfeldt in this country) to be the most rugged and convenient design. Samples can be introduced into the filter by cannula transfer through a serum cap, or by inverting the whole apparatus and attaching it directly onto a Schlenk tube containing the solution to be filtered. Removal of fine filtrates can be assisted by the use a filter aid such as Celite 545.

Purification Techniques

Recrystallization. The convenience of cannula techniques for the manipulation of solutions allows access to a wide variety of recrystallization techniques under inert atmosphere conditions. The general procedure in all recrystallizations is to extract a crude compound into a solvent with a polarity close to the minimum required to dissolve the material, filter the solution, and then reduce the polarity of the medium until the compound crystallizes out of solution. The product may then be collected by filtration, leaving small quantities of compounds with similar or higher solubilities in the solute mother liquor.

The reduction in medium polarity may be achieved in a number of ways, of which the simplest is to dissolve the crude product in hot solvent and allow the solution to cool. The filter stick filtration is particularly convenient for this, since it is straightforward to maintain the solution at any desired temperature during the filtration process.

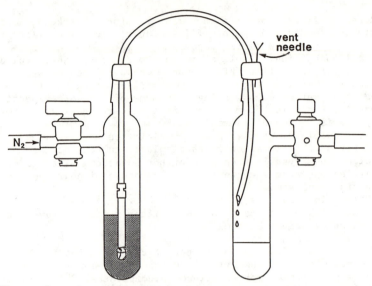

Figure 5.   Inert atmosphere filtration using filterstick.

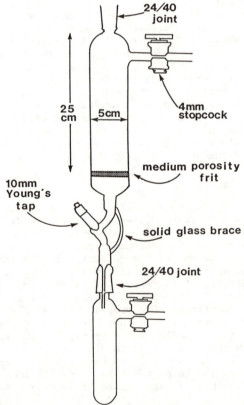

Figure 6.   Fritted glass filtration apparatus for Schlenk work.

Alternatively, a solution of crude product prepared at room temperature may be cooled to a low temperature. Filter stick filtration is again convenient for this procedure since it facilitates collection of the crystallized product at low temperature. It is common to cool the solvent to -78°C by slow addition of dry ice to an acetone cooling bath, but one convenient variation is to place the room temperature solution inside a Dewar flask containing some isopropanol in a low temperature freezer (-80°C). The solution will then cool over a period of 6 - 12 hrs to ~ -70°C.

A particularly convenient way of reducing the polarity of the nonaqueous solvent is to layer a less polar, lighter solvent over the filtered solution of the crude product as shown in Figure 7. Diffusion of the lighter solvent into the heavier solvent is surprisingly slow, and layer recrystallizations will typically need occassional agitation to ensure a complete mixing of the two layers over a convenient time scale of several days. Typical solvent combinations which can provide good recrystallization conditions include a layer of pentane over toluene (convenient for a nonpolar organometallic with significant solubility in aromatic solvents) or diethyl ether over tetrahydrofuran or dichloromethane.

A second approach to changing the composition of a mixture of two solvents is to use a less polar solvent which is also less volatile than the solvent in which the compound is initially dissolved. After the less polar solvent has been added to the filtered solution of the crude mixture, the mixture is concentrated under reduced pressure. The more volatile polar solvent is preferentially removed under these conditions, and the product will crystallize from solution. Typical solvent combinations which we have found to be valuable include benzene with 100-120°C ligroin, which can be conveniently used to crystallize neutral molecules which are soluble to aromatic solvents but not in hydrocarbons, and acetone with ethanol. Acetone will typically, for example, dissolve hexafluorophosphate salts of organometallic cations, which tend to be insoluble in the less volatile ethanol.

Compounds which are particularly difficult to crystallize, or which are exceptionally sensitive, may often be successfully crystallized by a vapor diffusion technique. The technique uses the apparatus shown in Figure 8, and in a typical recrystallization a solution of the compound in a polar solvent is placed in one of the two flasks, while a volatile, less polar solvent is placed in the second flask. The whole system is then placed under reduced pressure. Diffusion of the less polar solvent into the solution may be controlled by the pressure within the apparatus, or by introducing a temperature differential by cooling or warming one of the flasks as convenient.

Sublimation. Sublimation is less commonly used than it used to be, but can still be a valuable technique for the purification of small quantities of air sensitive organometallics. Its main disadvantages are that it is difficult to scale sublimations up to multiple gram quantities, and that thermal degradation typically results in significant loss of material during purification. Sublimation can be carried out using any of the commercially available sublimation

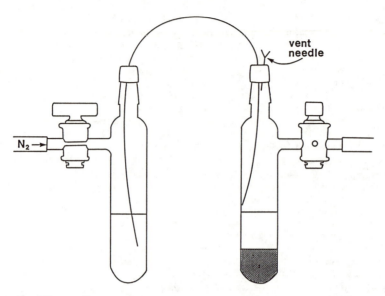

Figure 7.   Setting up a layer recrystallization.

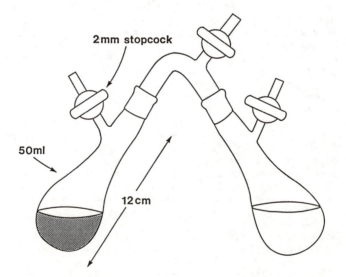

Figure 8.   Vapor diffusion recrystallization in progress.

apparatus, but a useful variation is to use a cold finger loaded with dry ice or liquid nitrogen (or water cooled), which will fit into the top of a small 50 mL Schlenk tube (see Figure 9).

Chromatography. Although column chromatography does not have the universal utility in inorganic chemistry that it does in organic chemistry, it can be a valuable separation technique for less sensitive organometallic compounds. Cannula techniques do allow column chromatography to be carried out under anaerobic conditions without requiring recourse to expensive specialized equipment. A typical anaerobic chromatography setup is illustrated in Figure 10. A standard chromatography column may be used, and the column is packed on the open bench. After the column has been packed it is sealed with a serum cap and the air gap is flushed with nitrogen through a needle. The packing material is degassed by passing 3 or 4 column volumes of degassed solvent through the column under a slight overpressure of nitrogen. The liquid level is run down until it is just above the adsorbent level, and the solution containing the compound to be purified is carefully transferred by cannula on to the top of the column. The column width is chosen so that the material can be added in a narrow band at the top, and separations are most successful when the column length is ~ 10 times the column width.

The column is eluted with degassed solvents transferred on to the top of the column through stainless steel cannulae, and a convenient feature of cannula chromatography is that it is particularly easy to set up a continuously variable polarity gradient in the elution medium as shown in Figure 11. Eluted bands may be readily collected under anaerobic conditons if the tip of the column is introduced into a nitrogen filled Schlenk through a bored serum cap. The only experimental complication is that it is essential to have a venting needle in the top of the collecting flask to avoid back pressure through the column.

Anaerobic manipulation of spectroscopic and analytical samples. Anaerobic manipulation of spectroscopic and analytical samples requires few additional techniques beyond those that have already been described. Samples which are to be shipped to other laboratories (for example, for combustion analysis) are most conveniently shipped under vacuum in sealed glass ampules, and Figure 12 shows the very simple apparatus used to prepare such sealed samples. Solid samples are typically loaded in a glove box or glove bag into a short section of 8 mm glass tubing which has been sealed at one end, the tube is then attached to a 6 mm high vacuum stopcock, and the sample is sealed off with a 3 point seal under vacuum. The vacuum provided by a rotary pump is sufficient for this manipulation, but it is often convenient to have one line equipped with an oil diffusion pump to provide somewhat higher vacuum conditons which will more efficiently remove traces of hydrocarbon solvents from analytical samples.

Solid state infra red spectra of air sensitive materials are more conveniently recorded as Nujol or hexachlorobutadiene mulls than as KBr pellets, primarily because the Nujol mull provides a protective coating for the solid material during sample preparation.

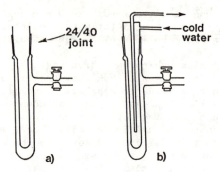

Figure 9.   Cold fingers for small scale sublimation.

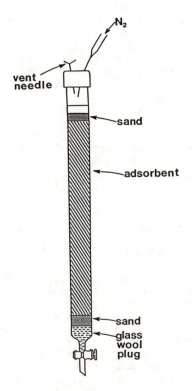

Figure 10.   Anaerobic column chromatography.

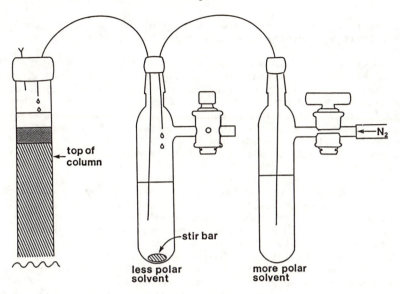

Figure 11. Establishing polarity gradient for chromatography.

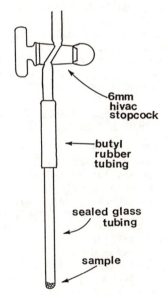

Figure 12. Apparatus for the preparation of sealed samples.

This significantly inhibits the kinetics of decomposition, and surprisingly air sensitive materials can be mulled on the open bench.

Solution infra red spectra can be recorded using standard cells, provided small serum caps are used to seal the cells from the atmosphere and the cells are flushed with degassed solvent before the solution is added through a fine (22 gauge) cannula. We have found little advantage in amalgam sealed cells, and prefer to use standard solution cells with teflon spacers. We find that these are adequately air tight provided the plates are kept in good condition by frequent polishing. With exceptionally sensitive solutions it is often convenient to scrub the cell with a sample of the solution to be examined before the spectrum is recorded.

The preparation of samples for NMR spectroscopy under anaerobic conditions used to be particularly time consuming. Two recent advances have, however, considerably simplified the preparation of such samples. One is the advent of reusable NMR tubes with an effective vacuum seal. These are manufactured by J. Young of London, and are available in this country through Brunfeldt, and we have found these relatively modestly priced NMR tubes to be satisfactory for all except the most sensitive materials. The second advance is the common availability of inert atmosphere glove boxes for filling tubes, but it is possible to load even the most sensitive solutions into an NMR tube without using a glove box by means of the simple adapter shown in Figure 13. This essentially converts the NMR tube into a micro Schlenk tube, and after the sample has been transferred into the tube via cannula the sample may be frozen in liquid nitrogen and the tube sealed under vacuum. The only difficulty is that the sample <u>must</u> be frozen from the bottom of the NMR tube to the top (for example by repeatedly dipping the NMR tube into and out of the liquid nitrogen) and <u>must</u> be thawed in the reverse direction. **CAUTION:** <u>If thawing begins at the bottom of the NMR tube it is not unusual for the tube to explode</u> as the phase change increases the volume.

Single crystal x-ray diffraction studies are frequently critical to the completion of organometallic research projects, and mounting crystals which are extremely air sensitive probably demands more dexterity than any other single experimental operation in a synthetic inorganic laboratory. Many groups solve this problem by using specially designed small glove boxes, but it is possible to mount crystals of even the most air sensitive materials by means of the inexpensive Schlenk tube apparatus shown on Figure 14.

A Schlenk tube is loaded with a sample of the crystals to be mounted, and with commercially available Lindemann tubes (Charles Supper & Company). The serum cap on the Schlenk tube is replaced with one which has been drilled to take a 8 mm glass tube, and crystals are then manipulated within the Schlenk tube by means of a fine-glass fiber, drawn from glass tubing, which is passed down this guide tube. The guide tube serves to steady the glass fiber, and to minimize back diffusion of air into the Schlenk tube. Manipulation is carried out under a very gentle countercurrent of nitrogen, and crystals are picked up by means of a small quantity of silicone grease on the tip of the glass fiber. Individual crystals are

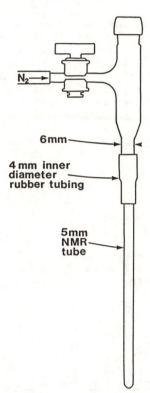

Figure 13. Adapter for the preparation of sealed NMR samples.

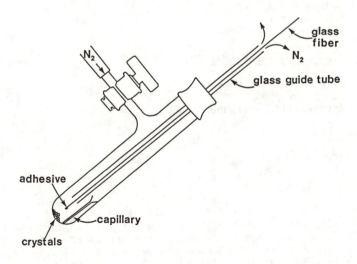

Figure 14. Schlenk tube equipped for the manipulation and mounting of air sensitive crystals.

placed into Lindemann tubes, and when 3 or 4 promising crystals have been mounted the serum cap is removed and the Lindemann tubes temporarily blocked with silicone grease. They may then be removed and sealed off with a very small gas flame. Convenient sizes of Lindemann tube for this operation are 0.3 mm or 0.5 mm, but any size appropriate to the crystal may be used. The preferences of crystallographers vary, but we have frequently found that the small quantity of silicone grease transferred with the crystal to the Lindemann tube provides sufficient adhesion to hold crystals in place without further glues. It may, however, be desirable to use Duco cement or a five minute Epoxy resin cement to mount the crystals more permanently.

Suppliers

Krytox. Available directly from:

> E. I. DuPont
> Chemicals and Pigments Division
> Speciality Chemical Products Division
> Wilmington, Delaware   19898

Teflon Tubing.

> Daiger Scientific
> 159 West Kinzie St.
> Chicago, Illinois   60610

Observation Tubes and Stainless Steel Cannulae.

> Popper and Sons
> 300 Denton Avenue
> New Hyde Park
> New York, N.Y.   11040

Young's Taps and Reusable NMR Tubes

> R. J. Brunfeldt Company
> P. O. Box 2066
> Bartlesville, Oklahoma   74005

Lindemann Tubes

> Charles Supper and Co,
> 15 Tech Circle
> Tatick, MA   01760

Suba Seals. Size 33 suba seals fit male 24/40 joints, size 49 fit female 24/40 joints, and size 29 provide a tight fit for female 14/20. Smaller sizes are often useful for specialized apparatus such as solution infrared cells. Suba seals are available from two suppliers in this country: Aldrich and Strem Chemicals.

Acknowledgments

We thank the National Science Foundation and the Office for Naval Research for financial support of our research program. We would also like to acknowledge the many colleagues and students who have contributed to the development of these techniques. We would particularly like to acknowledge our debt to Dr. Malcolm Green – many of the cannula techniques described are variations on techniques which were originally developed in his laboratories at Oxford University.

Literature Cited

1. Shriver, D. F., The Manipulation of Air Sensitive Compounds; McGraw-Hill: New York, 1969.

2. Burlitch, J. M., Booklet No. 570; Ace Glass: Vineland, N.J.

3. Kramer, G. W.; Levy, A. B.; Midland, M. M., In Organic Synthesis via Boranes; Brown, H. C., Ed.; Wiley: New York, 1975; Chapter 9.

RECEIVED July 31, 1987

High-Pressure Liquid Chromatographic Analysis of Air- and Water-Sensitive Compounds Using Gel Permeation Chromatography

James F. Burlitch and Richard C. Winterton

Baker Laboratory, Department of Chemistry, Cornell University, Ithaca, NY 14853-1301

The HPLC analysis of a mixture of phenylchlorosilanes using gel permeation columns is described. The method is applicable to a variety of organometallics or other compounds that are sensitive to air and water.

In the course of analyzing mixtures of halogenated organometallic compounds, we developed a method for the analysis of various polar organometallics that are sensitive to small amounts of air and water. That method proved to be versatile in several types of analyses. We describe here its application to the separation of several organo-chlorosilanes. HPLC has been used for the analysis of a variety of organometallic compounds (1, and references cited therein). In many cases, however, the degree of air and/or water sensitivity and any special handling techniques were not described.

Gel permeation columns were selected for this task because (1) the highly polar nature of the analytes precludes adsorption chroma-tography (the chlorosilanes and their tin analogues are too strongly adsorbed) and (2) because the required removal of water from solvents would have much less effect on retention times.

The mobile phase, toluene, was first deaerated by bubbling nitrogen through it while being stirred magnetically for several hours. This operation was carried out in the original bottle which had been equipped with a cap having stainless steel tubes glued into it for gas inlet, gas exit and solvent delivery. This process differs little from the usual practice of deaeration of the mobile phase except that the purge gas, after leaving the bottle, was led to an oil bubbler to prevent back diffusion of air or moisure (Fig. 1).

To reduce the water content to a tolerable level (< 1ppm), the mobile phase was passed through a 2.5 x 60 cm stainless steel column con-taining activated Linde type 3A Molecular Sieves™ prior to entering the HPLC pump. The Molecular Sieves (in bead form, held in place by

glass wool) had been activated at 200 - 250 °C under vacauum in a
tube furnace and when cool, were vented directly to the solvent
through a three-way Teflon™ valve. We found that if the column was
vented to nitrogen then subsequently filled with solvent , it was
extremely difficult to remove all of the nitrogen; such gas would
invariably cause malfunction of the detector. The drying column
reduced the water content of the toluene from 60 ppm to less than 0.5
ppm at a flow of 4 mL/min, as measured by a solid state moisture
detector (Panametrics, Model 1000). Little change in effectiveness
was seen during the use of several gallons of solvent.

Phenylchlorosilanes, Ph_3SiCl, Ph_2SiCl_2 and $PhSiCl_3$ (Alfa/Ventron)
were analyzed on four Waters Associates 100 Å μ-Styragel™ columns
arranged in series and equipped with a Model U6K injector. Solutions
of the silanes in toluene were transferred from septa-sealed vials
(Pierce Chemical Co.) with a 25 μL HPLC syringe. A typical chro-
matogram is shown in Fig. 2. Reproducibility of peak heights was
excellent in spite of the extreme water sensitivity of the compounds.
The order of elution, viz. that given above, was consistent with a

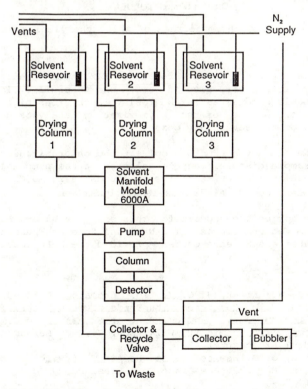

Figure 1. Schematic diagram of components for solvent purification and sample manipulation.

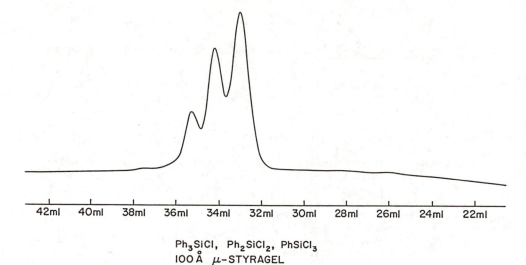

Ph₃SiCl, Ph₂SiCl₂, PhSiCl₃
100 Å μ-STYRAGEL
0.50 ml/min - TOLUENE

Figure 2.   HPLC chromatograms of three phenylchlorosilanes.

size separation process. The broadness of the peaks, suggested that
some other factors were also at work; one possibility is complexation at Lewis base sites on the resin. Such sites might be
byproducts or residues of the polymerization catalysts used in the
production of the stationary phase.

Although we did not attempt to collect the chlorosilanes, we found
that the sample disposal/recycle selector on the pump inlet manifold,
made collection of such sensitive compounds as Zn[Mo(CO)₃Cp]2 (2)
into Ace Glass Co. No-Air™ vessels (3) quite easy.

We greatly appreciate support for this study by Waters Associates and
advice from Kenneth Conroe. The National Science Foundation, via the
Cornell Materials Science Center, provided the HPLC instrument.

Literature Cited

1. Gast, C. H.; Kraak, J. C. J. Liq. Chromatog. 1981, 4, 765.
2. Burlitch, J. M. Compounds with Bonds between Transition Metals
    and Either Mercury, Cadmium, Zinc or Magnesium in Comprehensive
    Organometallic Chemistry, Wilkinson, G., ed. Pergamon Press,
    Oxford, 1982, chapter 42.
3. Burlitch, J. M. How to Use Ace No-Air Glassware; Technical
    Bulletin No. 570, Ace Glass Co.: Vineland, N. J.

RECEIVED September 1, 1987

Low-Temperature Chromatography Column for Routine Use in Organometallic Separations

Robert C. Buck and Maurice S. Brookhart

Department of Chemistry, University of North Carolina, Chapel Hill, NC 27514

We report here a design for a low-temperature chromatography column (Figure 1) which is easily constructed and used for the purification and/or separation of sensitive organometallic compounds which decompose or react when chromatographed at room temperature. The two-jacket design includes an outer vacuum jacket which allows for excellent temperature control and prevents icing while the column temperature is maintained by circulating a cooled solution of dry methanol through the inner jacket. Examples of compounds purified and/or separated include iron, manganese and cobalt organometallic complexes.

Column chromatography of sensitive organometallic compounds generally requires both solvents and supports which have been rigorously degassed. However, despite these precautions and the use of the most inert supports available, many organometallic compounds decompose or react on the column. We describe here a design for a low-temperature chromatography column which is easily constructed and readily set up for routine use. We have been successful in utilizing low-temperature chromatography to achieve the separation and purification of several organometallic compounds which otherwise decompose or react when chromatography is attempted at ambient temperature.

Although low-temperature chromatography is not new, our column design (Figure 1) offers the useful feature of an outer evacuated jacket which provides for excellent column temperature control and prevents fogging and ice build-up, allowing colored organometallic compounds to be viewed as elution proceeds. The column is constructed of heavy-wall glass and consists of a standard inert-atmosphere column surrounded by two jackets, an inner jacket through which a coolant is circulated and an outer vacuum jacket which completely encloses the inner jacket. The specific joints used and actual column dimensions are detailed in Figure 1. The site which can ice up is the length of tubing between the base of the vacuum jacket and the 4mm Teflon stopcock. When wrapped with glass wool or a paper towel this area is sufficiently insulated to prevent icing.

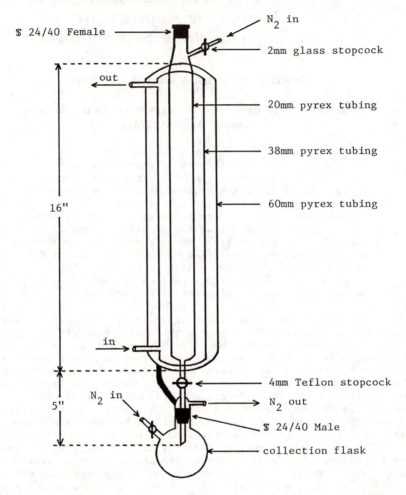

Figure 1.   Schematic of Low-Temperature Chromatography Column.

Another potential danger is the implosion of the outer vacuum jacket. For safety, we wrap strips of transparent tape in a grid around the column and place a Plexiglas shield in front of it when in use.

The column is packed at room temperature then slowly cooled to the desired temperature with a dry methanol solution. We have used two methods to cool the methanol:

(1) Three liters of dry methanol cooled with dry ice are placed in a 4 quart PVC foam bucket. A Little Giant Pump (Model #540052, fitted with a wire gauze screen to prevent impellar breakage from small pieces of dry ice) is used for circulating the methanol through mininal lengths of heavy-wall butyl rubber tubing (1/4" bore, 1/8" wall thickness). Column temperatures of ca. $-75°C$ can be achieved.

(2) A NESLAB ULT-80DD Refrigerated Circulating Bath equipped with a "z-suction" pump. The 15 liter bath is filled with dry methanol which is cooled to the desired temperature and circulated through minimal lengths of butyl rubber tubing (same as above) wrapped with foam pipe insulation. The bath temperature is regulated and monitored by a digital readout on the unit.

Examples of compounds which we have successfully purified and/or separated include: the two diastereomers of $(n^5-C_5H_5)(CO)(PEt_3)FeCH(OMe)Ph$ on Act. II-III basic alumina at $-50°C$ with Hexanes; the two diastereomers of $(n^5-C_5H_5)(CO)(PEt_3)FeCH(OMe)Me$ on Act. II-III basic alumina at $-75°C$ using 8:1 Hexanes/Ethyl Acetate (these compounds readily lose MeOH to form the vinyl derivatives upon chromatography at $25°C$); $(n^6-toluene)(CO)_2MnEt$ on neutral alumina at $-15°C$ with 1:1 Hexanes/Et$_2$O; and $(n^5-C_5Me_5)(PMe_3)Co(Me)_2$ and $(n^5-C_5Me_5)(PMe_3)Co(Me)(I)$ on florisil at $-75°C$ using Methylene Chloride.

RECEIVED July 31, 1987

Chapter 2: Application 3

IR Radiation as a Heat Source in Vacuum Sublimation

Guillaume C. Pool and Jan H. Teuben

Department of Inorganic Chemistry, Nijenborgh 16, 9747 AG Groningen, Netherlands

Vacuum sublimation is a convenient purification technique, frequently used in synthetic organometallic chemistry. There are many types of sublimation flasks, varying chiefly in degree of sophistication of design. The main practical problem with vacuum sublimation is the inefficient transfer of heat from heat source to the sample being sublimed. Virtually all set ups use an oil- or metal bath and rely on conduction as the main mechanism of heat transfer. Under the vacuum conditions of the usual sublimation ($10^{-1} - 10^{-3}$ mm Hg) both convection and conduction of heat are very inefficient and sublimation is a very slow process.

In our group we have developped an alternative method by using an infrared heat lamp which gives very efficient heat transfer and speeds up sublimation dramatically. For instance the rather involatile Group 5 compounds Cp_2MCl_2 (M = Nb, Ta), which sublime at ca. 260°C at 10^{-3} mm Hg are sublimed easily by infrared radiation at a rate of 3 g per hour (metal bath heating requires 24 h or more for the same amount). We routinely use the IR radiation set up for virtually all starting organometallics e.g. Cp_2Mg, CpTl, Cp_2V, Cp_2TiCl, $CpTiCl_3$, $Cp*MCl_3$ ($Cp* = C_5Me_5$, M = Ti, Zr, Hf) that we use as well as for new compounds. Only in the case of thermally labile compounds (e.g. $Cp*TiMe_3$) is this technique less satisfactory.

Technique

We use a sublimation set-up as sketched below (Figure 1). A Philips IR lamp (375 W/ 220 V) connected to a variable resistor is sufficient for most sublimations. The lamp is protected from the sublimation flask by a metal gauze screen. The lamp is placed in a metal cylinder with reflecting walls and the sublimation flask is inserted through a conical reflector which can be adjusted to fit the flask. Dimensions are not critical and there is ample opportunity for improvisation e.g. use of aluminum foil as a disposable reflector.

Controlling the temperature within a narrow range is rather difficult. Trial and error will establish the temperature achieved for a certain type of sublimation flask at a specified applied voltage. We have determined the temperature in the center of a sublima-

0097–6156/87/0357–0030$06.00/0

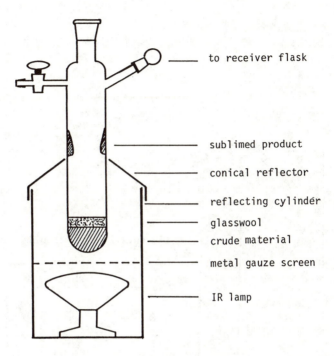

to receiver flask

sublimed product

conical reflector

reflecting cylinder

glasswool

crude material

metal gauze screen

IR lamp

Figure 1. Sublimation set-up.

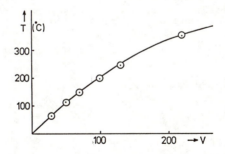

Figure 2. Plot of temperature vs. voltage.

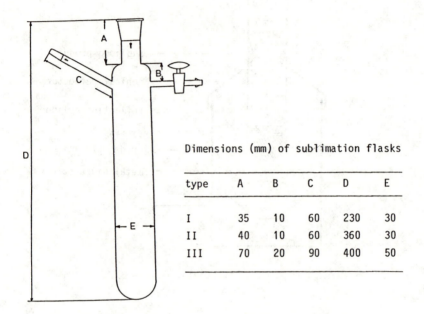

Dimensions (mm) of sublimation flasks

type	A	B	C	D	E
I	35	10	60	230	30
II	40	10	60	360	30
III	70	20	90	400	50

Figure 3. Sublimation flasks.

tion flask filled with (4 cm diameter, 5 cm height) respectively Al_2O_3 (white) and carbon powder (black) under sublimation conditions ($p < 10^{-3}$ mm Hg). In both cases the maximum temperature reached was the same at a particular voltage. In all cases this maximum temperature was reached within 1 h, the black carbon powder reaching maximum temperature considerably faster.

It is observed that the temperature achieved strongly depends on both the diameter of the sublimation flask, and on the fit of the flask in the conical reflector top entrance.

A typical T/V plot is given below (Figure 2). The temperature range is clearly larger than accessible by oil or metal bath heating.

Sublimation Flask

For simplicity, we have chosen long Schlenk tubes (Figure 3), without the conventional cold finger condenser. Deposition of crystals takes place on the cold walls of the sublimation flask 1-2 cm above the conical reflector.

We find it quite easy to isolate the sublimed product without the use of a drybox. The isolation procedure is as follows. The sublimer is positioned horizontally, the stopper at the top is removed and with N_2 (Ar) flowing out of the flask the crystals are scraped from the walls and powdered using a stainless steel spatula. The product is then transfered to a storage flask connected through the side arm.

The sublimer should not be overloaded. We generally avoid having more than 3-4 cm of crude material at the bottom of the flask. The residue is kept in place by a loosely packed glasswool plug.

Advantage

The method described here is cheaper and safer than the conventional methods (silicon oil or Woods' metal bath). In addition it is much cleaner since no oil or metal stick to the walls of the sublimation flask after it is removed from the sublimer. The main advantages are the high sublimation rates and higher temperatures that can be reached relative to the conventional techniques.

RECEIVED July 31, 1987

Chapter 3

Techniques in the Handling of Highly Reduced Organometallics

John E. Ellis

Department of Chemistry, University of Minnesota, Minneapolis, MN 55455

Procedures used and developed in our laboratory for the
synthesis, isolation and characterization of highly air
sensitive organometallic compounds are reviewed.
Details of the inert atmosphere purification system,
vacuum line and specialized glassware employed in these
operations are also presented.

The objective of this account is to describe some of the apparatus
and procedures used in this laboratory for the synthesis, isolation
and purification of highly moisture and oxygen sensitive organometallic compounds. Special attention will be given to the inert
gas purification system, vacuum line, transfer methods and specialized glassware and associated apparatus employed in these operations.
This treatment is not intended to be a complete guide to the synthesis and handling of air sensitive materials, but it is hoped that
the details and observations presented herein will be especially useful to novices in this area and those who have not had the opportunity to visit laboratories in which such chemistry is routinely
carried out. Undoubtedly, the best general compilation of such procedures and techniques is the recently published book by D.F. Shriver
and M.A. Drezdzon.(1) Several of the diagrams of apparatus presented herein represent rather simple modifications of glassware
initially observed in other laboratories, journal articles, especially Journal of Chemical Education, Inorganic Syntheses, catalogs
of scientific glassware (especially Ace Glass and Kontes Glass
Companies), and other sources.(2-12) However, unlike most drawings
in the literature, the ones presented herein are sufficiently
detailed to be sent directly to the glassblower for fabrication.
Hopefully this useful feature will warrant their inclusion.
   Any laboratory involved in organometallic research should have
available a variety of methods for the handling of materials of
diverse physical and chemical properties. These include glove boxes,
high and medium vacuum lines and associated glassware and apparatus.

0097–6156/87/0357–0034$09.00/0
© 1987 American Chemical Society

Although some organometallic research groups conduct practically all of their synthetic operations in highly efficient (and expensive!) commercial glove boxes, we prefer to use a glove box mainly for the preparation of samples for spectroscopic and elemental analysis, transfer and weighing of solid reactants or products and storage of samples. In general, it has been our experience that organometallics survive better in the atmosphere supplied by a vacuum-inert gas manifold than that of a glove box, even after the dry train charge of the glove box (Vacuum Atmospheres Corp.) has been regenerated. For this reason highly sensitive solids should be stored in sealed glass ampoules, Schlenk tubes or other vessels which are impermeable to air, rather than screw capped vials or bottles in a glove box. Indeed, many of the substances that members of our group routinely handle appear to be more efficient in removing oxygen or moisture from an "inert" atmosphere than the purification train of our glove box. For this reason, materials to be submitted for elemental analysis are handled with particular care. After the samples are placed in ampoules, which are sealed to a modified Schlenk tube, the ampoules are evacuated and refilled three or four times with inert gas supplied from our vacuum line. They are then sealed off, inspected after a day with a hand lens or microscope for sample deterioration and sent out for analysis if everything looks satisfactory. Also, in so far as possible, samples submitted for analysis are handled only in solvent free glove boxes.

   Other groups have emphasized the importance of using high vacuum (10^{-3} to 10^{-6} torr) lines in organometallic research. While this technique is almost mandatory for the handling of milligram quantities of highly air sensitive materials, especially when they are present in dilute solutions or of high molecular weight, physical studies where traces of oxidation products cannot be tolerated, quantitative transfers of gases or the distillation and sublimation of thermally sensitive materials of low volatility, it can be an inefficient and overly rigorous way to do organometallic chemistry. For these reasons, most of the operations in this laboratory are conducted on the bench with a double manifold medium vacuum (10^{-1} to 10^{-3} torr) line such as the one described later in this article. This is a flexible system which is suitable for most operations one will encounter in organometallic or organic syntheses and requires little maintenance after assembly. Provided reactions are conducted with care on a reasonable scale (<u>ca.</u> > 25 mmole), there should be minimal problems with reactant or product decomposition by air. Principal advantages of this technique over those involving glove boxes or high vacuum lines include relatively small space requirements, low cost, excellent flexibility and ease of manipulations, and facile transfers of solutions by syringe or cannulae. A significant disadvantage, however, is that a person performing an operation with a vacuum line on an open bench or in a hood often needs to be more highly trained, meticulous and acutely aware of potential dangers than the individual who is carrying out roughly the same task in a glove box, where mistakes and/or equipment failure are less likely to result in personal injury. One of the most common mistakes made with a vacuum line is for one to unwittingly open a tap on the line to a flask containing air while another vessel is being evacuated. If the latter flask contains a highly oxygen sensitive

material an explosion could result. This is particularly a problem
with highly reduced metal carbonyls which will often explode when
exposed to even small amounts of air in a closed system. A cardinal
rule to be observed when using any vacuum line is that only one
vessel should be open to the vacuum chamber at any given time. Any
contemplated violation of this rule must be weighed very carefully!
Many other hazards are associated with the use of vacuum lines which
are beyond the scope of this brief report. In this regard, part 2
of Shriver and Drezdzon (1) should be read soon and thoroughly if
you are a newcomer to the world of vacuum lines!

The Inert Atmosphere Purification System

Although some authors have recommended the use of high purity inert
gases without further purification (12) due to the real possibility
of introducing impurities during passage through a poorly designed or
contaminated "purification" system, an effective gas purification
train can be exceedingly important in reducing oxygen and moisture
to tolerable levels. Our research has required the use of such a
system. One can purchase "research grade" nitrogen or argon which
contains less than 10 ppm total impurities, but it is prohibitively
expensive for normal synthetic work. Prepurified grades of nitrogen
or argon (99.998 mole % pure) are reported to contain less than 3 ppm
H_2O and 10 ppm or 3 ppm O_2, respectively (13), which are acceptable
levels for many purposes in organometallic chemistry. If the inert
gas in cylinders were always of uniform and specified purity, one
could undoubtedly do much organometallic chemistry without a gas
purification system, but it has been our experience that this is
generally not the case. Indeed, prepurified grades of these gases –
at least from our suppliers – or even "house nitrogen" which is
obtained from the evaporation of liquid nitrogen and circulated
throughout Departmental laboratories – will often cause hydrocarbon
solutions of $TiCl_4$ or $Zn(C_2H_5)_2$ to smoke, indicating that unaccep-
table levels of moisture or oxygen, are present. If
the system is properly constructed (i.e., free of leaks) of materials
which are impermeable to air and is maintained continuously at positive pressure, improved or at least reproducible results in organometallic research should be obtained, unless the process is catalyzed
or promoted by adventitious oxygen or moisture!
   A schematic drawing of the inert atmosphere purification system
used in our laboratory is shown in Figure 1. Features of this system
are as follows:
1. All components are glass or metal except Teflon ferrules and
    stopcock plugs.
2. Swagelok (or Gyrolok) fittings and copper tubing provide flexi-
    bility and minimize the use of fragile glass connections.
3. Teflon front ferrules in Swagelok fittings provide leak free
    glass to copper seals and eliminate the need for more expensive
    and fragile Kovar or other metal to glass seals.
4. The catalyst is maintained at about $150^{\circ}C$ to increase its oxygen
    absorption capacity (about 5x that at $25^{\circ}C$).
5. Wherever possible, greaseless connections are incorporated to
    minimize maintenance. With normal use and source gas purity,
    the BASF catalyst and molecular sieves need to be regenerated
    approximately once every 5-6 years.

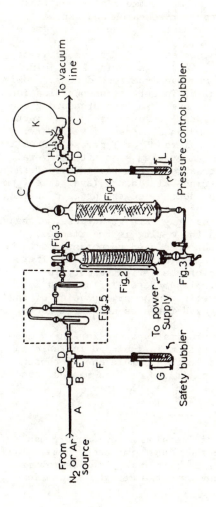

IDENTIFICATION OF COMPONENTS AND DIMENSIONS

A 1/4 IN OD COPPER TUBING

B SWAGELOK UNION, 1/4 X 3/8 IN

C 3/8 IN OD COPPER TUBING

D SWAGELOK T, 3/8 IN

E 75 MM, 3/8 IN OD GLASS TUBING

F 80 CM, 8 MM OD CAPILLARY TUBING

G 20 CM COLUMN OF MERCURY

H SWAGELOK UNION, 3/8 IN

I 5 CM, 3/8 IN OD GLASS TUBING

J 15 MM ID O-RING SEAL JOINT

K 12 LITER SURGE FLASK

L 12 CM COLUMN OF MERCURY

Figure 1.   Schematic view of the inert gas purification system.

6. The system is maintained at about 12 cm Hg positive pressure.
   Incorporation of a 10 or 12 liter surge flask permits the filling
   of up to a 2 liter vessel without creating a negative pressure
   (VERBOTEN!) or having to greatly increase the flow rate of gas
   through the purifier. The optional Nujol bubbler assembly provides a qualitative measure of flow rate.

Although many groups now use MnO-vermiculite or silica gel as an
oxygen scavenger, we find that the standard commercially available
BASF catalyst works satisfactorily for our research and can be
regenerated at substantially lower temperatures than the manganese
based material (see ref. 1 pp. 74-80).

   Detailed diagrams of the components of the inert gas purification system are shown in Figures 2-5. Although o-ring seal joints
are specified for many connections, we have experienced considerably
more difficulty (and occasional anguish) in setting up purification
trains with these rather inflexible joints than with standard ground
spherical (i.e., ball and socket) joints. Unfortunately, the latter
must be carefully checked for warps and hand-lapped if necessary;
however, if they are lubricated with Apiezon H or T grease and well
matched, channelling is generally not a problem even after several
years as long as the internal gas pressure is not too great. Perhaps
the best solution would be to use o-ring ball joints (available from
Kontes or Ace Glass Co.). These are about twice the price of o-ring
seal joints but have the flexibility of spherical joints and the
advantage of o-ring seal joints in that they are greaseless, although
a light coating of Apiezon H or T on the o-ring is recommended.
Another desirable substitution for our specified components would be
to use graphite or Vespel front ferrules (available from suppliers
of chromatography equipment) instead of the Teflon variety for
glass-metal tubing connections. Ferrules made of Teflon tend to
cold-flow which often necessitates retightening or replacing of the
ferrule after some time. Graphite based ferrules are easily deformed
but do not tend to "cold flow".

   A summary of the sources and approximate prices for the components of the inert gas purification system is included in Table I.
Even at 1987 prices, one should be able to assemble one of these
systems for less than $1000. Although it is not recommended, one
purification line can serve more than one vacuum line.

Vacuum-Inert Atmosphere Manifold

Figure 6 depicts a simplified five tap version of our double manifold vacuum line. This assembly is shown as being about 5 feet long
but can be made considerably shorter as a four tap line and/or if
only one cold trap is used.† In many ways the design is superior

---

†The line is a modified version of one Stan Wreford first assembled
as a graduate student in Alan Davison's group at M.I.T. Stan
apparently borrowed the design from Hans Brintzinger's lab at the
University of Michigan, where he did undergraduate research.

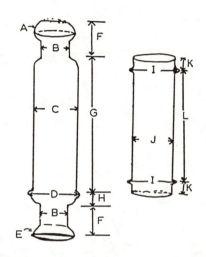

IDENTIFICATION OF COMPONENTS AND DIMENSIONS

A	S.J. 50/30 BALL JOINT	G	70 CM
B	35 MM OD	H	25 MM
C	60 MM OD	I	80-85 CM OD FLANGE
D	70-75 MM OD FLANGE	J	70 MM OD
E	30 MM ID ORSJ	K	50 MM
F	75 MM	L	60 MM

NOTE: ORSJ = O-RING SEAL JOINT

Figure 2.   Oxygen removal tower and outer column for thermal insulation.

Top Connection

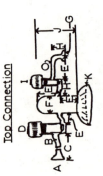

Bottom Connection

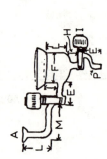

IDENTIFICATION OF COMPONENTS AND GLASS TUBING DIMENSIONS

A	15 MM ID ORSJ	F	35 MM OD	L	75 MM
B	15 MM OD	G	9 MM ID ORSJ	M	6 CM
C	20 CM	H	25 MM	N	30 MM ID ORSJ
D	KONTES K826500, 0-8 MM	I	KONTES K826500, 0-4 MM	O	10 MM OD
E	5 CM	J	10 CM	P	8 MM OD
		K	S.J. 50/30 SOCKET		

NOTE: ORSJ = O-RING SEAL JOINT

Figure 3. Top and bottom component of oxygen removal tower.

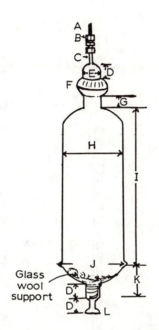

IDENTIFICATION OF COMPONENTS AND GLASS TUBING DIMENSIONS

A	3/8 IN COPPER TUBING	G	75 MM
B	SWAGELOK UNION, 3/8 IN	H	90 MM OD
C	50 MM; 3/8 IN OD GLASS	I	60 CM
D	50 MM	J	100-110 MM OD FLANGE
E	60 MM OD	K	10 CM
F	S.J. 75/50 SPHERICAL JOINTS. GREASE WITH APIEZON H OR T	L	15 MM ID O-RING SEAL JOINT

Figure 4. Molecular sieve tower.

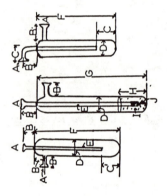

IDENTIFICATION OF GLASS COMPONENTS AND DIMENSIONS

A 15 MM ID O-RING SEAL JOINT

B 25 MM

C 50 MM

D 40 MM OD

E 10 MM OD

F 25 CM

G 40 CM

H 10 CM

I EXTRA COARSE GLASS DISPERSION TUBE

J 10 MM

Figure 5. Nujol bubbler assembly of the gas purification system.

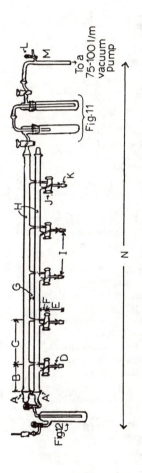

IDENTIFICATION OF GLASS COMPONENTS AND DIMENSIONS

A S.J. 35/20 SPHERICAL JOINTS

B 75 MM

C 150 MM

D GLASS SPURS, 10 MM ABOVE JOINT

E 12 CM

F 40 MM

G 25 MM

H 28 MM OD

I 150 MM

J 4 MM, 3 WAY HIGH VACUUM, HOLLOW PLUG STOPCOCKS

K S.T. 14/35 INNER JOINT

L KONTES K826500, 0-4 MM

M 25 MM OD

N CA. 150 CM TOTAL LENGTH

Figure 6. Double manifold vacuum line excluding manometers.

Table I. Summary of Sources and Approximate Cost for
Components of the Inert Gas Purification System

1. Swagelok Union, 1/4 x 3/8 Cat. No. B600-6-4 @ 2.65	Crawford Fitting Co. 29500 Solon Road Solon, Ohio 44139	2.65
2. Swagelok T, 3/8 x 3/8 x 3/8 Cat. No. B600-3 @ 6.05	Crawford Fitting Co. 29500 Solon Road Solon, Ohio 44139	18.15
3. Swagelok Teflon front ferrule 3/8 in. T-603-1 @ 1.06	Crawford Fitting Co. 29500 Solon Road Solon, Ohio 44139	5.30
4. Swagelok Union 3/8 x 3/8 Cat. No. B600-6 @ 2.65	Crawford Fitting Co. 29500 Solon Road Solon, Ohio 44139	5.30
5. Catalyst Tower and Outer Column (Figure 2)	Local fabrication at Glass Technology Service (Univ. of Minnesota)	65.00
6. Top and Bottom Connection for Catalyst Tower (Figure 3)	Local fabrication at Glass Technology Service (Univ. of Minnesota)	183.00
7. BASF or R3-11 Catalyst (1 kg)	Chemical Dynamics P.O. Box 395 South Plainfield, NJ 07080 (201) 753-5000	42.00 (or 28.00/kg for 5 kg unit)
8. Nichrome or Chromel A Wire (50 ft., B+S 20 Gauge, <u>ca.</u> 0.65 ohm/ft.)	Any Lab. Apparatus Distributor such as Sargent Welch	15.00
9. Molecular Sieve Tower (Figure 4)	Local Fabrication	80.00
10. Molecular Sieves, 13X, 4-8 mesh (<u>ca.</u> 5 lb.)	Fisher Scientific	30.00
11. Surge Flask, 12 liter	Local fabrication	57.00
12. Optional Nujol Bubbler Assembly (Figure 5)	Local fabrication	92.00
13. Capillary Tubing, 8 mm OD, 2 x 4 ft.	Fisher Scientific	8.00

Not included in list is copper and glass tubing, two pressure relief
bubblers, needle valve for control of the source gas, gas regulator,
mercury, and power supply for Catalyst Tower.

TOTAL ESTIMATED COST (1984-1985 PRICES) OF SYSTEM: $700.00

to a commercially available unit from the Ace Glass Co. In particular, the presence of standard taper or, alternatively, o-ring ball connections on each tap permit the attachment of a reaction vessel directly to the vacuum line via a glass or metal adapter. This procedure minimizes contamination problems associated with the use of plastic (e.g. Tygon) or butyl rubber tubing. These types of tubing are not only permeable to oxygen and moisture but also absorb solvent vapors which may then combine with reactants or products.††
Flexible stainless steel tubing is extremely useful when the presence of plastic or rubber tubing is undesirable, yet considerable flexibility is necessary between the vacuum line and apparatus. The use of flexible SS tubing should be considered for any long term operations involving air sensitive materials which must be left open to a manometer, gaseous reactant or inert atmosphere line. For example, in Figure 7, a typical assembly for conducting a reaction under an atmosphere of carbon monoxide is shown. This apparatus has proven to be useful in the synthesis of extremely air sensitive materials such as $Ti(CO)_3(dmpe)_2$ or $C_5H_5Ti(CO)_4^-$ by low temperature reductive carbonylations conducted at near atmospheric pressure over a period of 10-12 hours.(16) Flexible SS tubing is also particularly useful for sublimations conducted under dynamic vacuum conditions. In this case Ultra-Torr fittings should be used.

   Figure 8 shows a cross-sectional view of the vacuum line which details the manometer design. Manometers are attached to each vacuum tap to permit facile handling of gaseous reactants or low boiling solvents such as liquid ammonia or dimethyl ether. Manometers are important for pressure measurements particularly during reduced pressure distillations. They are also extremely useful for checking evacuated apparatus for leaks and function as simple pressure relief valves for ordinary distillations or reactions in which gases are evolved. The design may be simplified by having one manometer serve two taps via a three way T bore spring loaded stopcock

††A qualitative idea of the permeability of butyl rubber (e.g., Fisher 14-168B) or Tygon vacuum tubing to air may be gained by initially filling a meter length of tubing with pure argon or nitrogen, leaving the contents undisturbed for 12 hours or more and then flushing the gas from the tubing into a vessel containing a highly air sensitive compound in solution (e.g., $Na-C_{10}H_8$ in tetrahydrofuran or chromous chloride in water). Invariably, enough air will have diffused through the tubing within this period to cause significant decomposition of the test material. This observation should serve as ample warning to those who use rubber or plastic connections between a vacuum-inert gas line and reaction apparatus involving highly air sensitive materials. The latter should either be closed to the tubing, if left for extended periods of time, or connected to the vacuum line by nonpermeable materials.

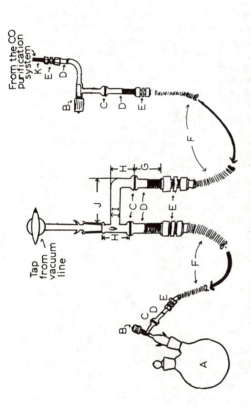

IDENTIFICATION OF COMPONENTS AND DIMENSIONS

A TWO NECK REACTION VESSEL EQUIPPED
   WITH A 90° STOPCOCK

B ROTAFLO 90° 0-3 MM TEFLON STOPCOCK

C 9 MM ID O-RING SEAL JOINT

D KOVAR-GLASS SEAL, 1/4 IN OD

E SWAGELOK UNION, 1/4 IN

F CAJON SS TUBING, 1/4 IN OD
   24 IN LENGTH

G 75 MM

H 50 MM

I 12 MM OD

J 60 MM

K 1/4 IN OD COPPER TUBING

Figure 7. Diagram showing the use of flexible stainless steel
tubing for conducting reactions under an atmosphere of carbon
monoxide or other gas.

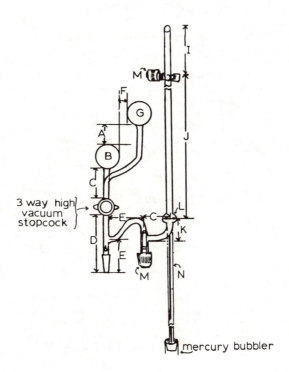

Figure 8. Cross-sectional view of the vacuum line showing a manometer. Continued on next page.

IDENTIFICATION OF

COMPONENTS AND DIMENSIONS

A   25 MM

B   INERT GAS CHAMBER

C   40 MM

D   12 CM

E   50 MM

F   6-12 MM

G   VACUUM CHAMBER

H   10 MM OD

I   10 CM

J   30 CM

K   65 MM

L   9 MM ID ORSJ

M   KONTES K826500, 0-4 MM

N   8 MM OD GLASS CAPILLARY

NOTES: ORSJ = O-RING SEAL JOINT

SEE FIGURE 10 FOR THE CORRECT
ORIENTATION OF THE 3 WAY
STOPCOCK

Figure 8.   Continued.

without greatly reducing the flexibility of the system. Another view
of the manometer is depicted in Figure 9, which also shows an
optional Nujol bubbler. This bubbler is sometimes useful for conducting operations near atmospheric pressure or qualitatively determining the flow rate of the inert gas through a tap before opening a
flask, etc. to the atmosphere with a countercurrent of nitrogen or
argon. Ideally, manometers should be fused directly to the vacuum
line to minimize the possibility of leaks, but in practice this is
inconvenient when the apparatus must be disassembled, unless a com-
petent glassblower is in residence. For this reason o-ring seal,
spherical or o-ring ball and socket joints are used in this connection. We have also tested the somewhat more expensive Solv-seal
joints (Fischer and Porter Co.) and find that they have no particular
advantage in this or other applications. The relative inflexibility
of o-ring seal joints is not a serious problem here. Although Kontes
Teflon stopcocks are specified for the manometers, any high vacuum
Teflon stopcock should be satisfactory. Figure 10 depicts possible
orientations of the stopcocks on the vacuum line. Other components
of the vacuum line are illustrated in Figures 11 and 12. The large
bore glass vacuum stopcocks shown in Figure 11 may be replaced by
high vacuum Teflon stopcocks provided their o-rings are fairly inert
to solvent vapors. O-rings made of remarkably inert Kalrez (Ace
Glass Co.) can be used but are extremely expensive. The molecular
sieve trap shown in Figure 12 is particularly useful in minimizing
contamination of the inert gas purification train by solvents or
volatile reagents. Such a trap, cooled with CO_2-iPrOH, is especially
important whenever reactions involving liquid ammonia are done on the
line. Ammonia vapor can easily back-diffuse into an unprotected
inert gas purification train and ruin much subsequent chemistry if
the problem is not detected in time. For this reason, all liquid
ammonia chemistry in this laboratory is generally done on only one
or two selected benches. Contamination of the "inert gas" by many
substances can seriously harm experiments conducted on the line and
must always be suspected when reactive materials decompose or slowly
die under mysterious circumstances. Such contamination can arise
from slow release of absorbed species from a molecular sieve column
(or trap!) and/or rubber or plastic tubing. The previously mentioned tendency for such tubings (especially Tygon) to strongly
absorb (and then release) solvent vapor and moisture may be more of
a problem than their inherent permeability to air when using these
materials as flexible connections on a vacuum line. Even vapor as
innocuous as tetrahydrofuran, one of the most commonly used solvents
in organometallic chemistry, can cause decomposition of strong
electrophiles such as $TiCl_4$ (which fumes strongly in nitrogen or
argon contaminated by tetrahydrofuran vapor) or $TaCl_5$ at room temperature.

   In closing this section, a consideration of some pros and cons
of using all glass stopcocks on a double manifold vacuum line is
appropriate, particularly in view of Wayda and Dye's recent design

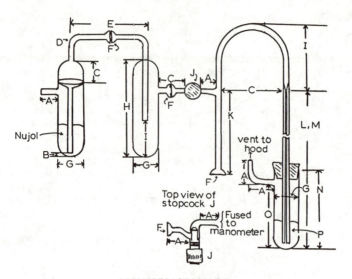

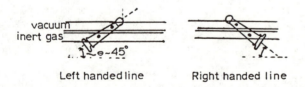

IDENTIFICATION OF

COMPONENTS AND DIMENSIONS

A	25 MM	I	10 CM	
B	20 MM	J	KONTES K826500, 0-4 MM	
C	50 MM			
D	8 MM OD	K	30 CM	
E	75 MM	L	80 CM	
F	9 MM ID ORSJ	M	8 MM OD. CAPILLARY	
G	25 MM OD	N	15 CM	
H	25 CM	O	13 CM	
		P	4 CM COLUMN OF MERCURY	

NOTE: ORSJ = O-RING SEAL JOINT

Figure 9.   Side view of a manometer and Nujol bubbler of the vacuum line.

Figure 10.   Top view of the manifold showing possible orientations of the stopcock.

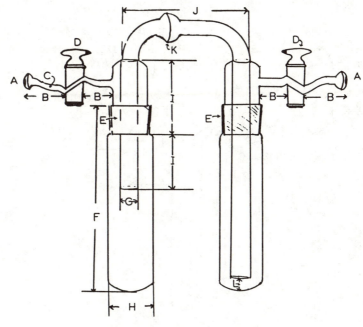

IDENTIFICATION OF COMPONENTS
AND DIMENSIONS

A   S.J. 35/20 BALL JOINT

B   50 MM

C   20 MM OD

D   10 MM OBLIQUE BORE, HOLLOW
    PLUG STOPCOCK

E   S.T. 45/50 GROUND GLASS
    JOINTS

F   30 CM

G   20 MM OD

H   50 MM OD

I   10 CM

J   18-20 CM

K   S.J. 35/20 SPHERICAL JOINTS

L   25 MM

NOTE: LARGE BORE TEFLON STOP-
COCKS MAY BE SUBSTITUTED FOR
COMPONENT D

Figure 11.   Cold trap assembly of the vacuum line.

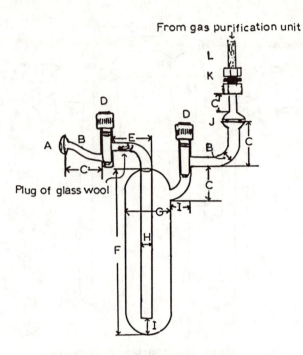

IDENTIFICATION OF COMPONENTS
AND DIMENSIONS

A   S.J. 35/20 BALL JOINT

B   15 MM OD

C   50 MM

D   KONTES K826500, 0-8 MM

E   75 MM

F   30 MM

G   50 MM OD

H   25 MM OD

I   25 MM

J   15 MM ID O-RING SEAL JOINTS

K   SWAGELOK UNION, 3/8 IN

L   3/8 IN COPPER TUBING

Figure 12.   Molecular sieve trap.

which uses only stopcocks with Teflon plugs.(14) Perhaps the principal advantage is using glass stopcocks relative to the Teflon variety is their resistance to scratching† and fouling by bumped solutions, volatile solids, etc. This latter property is especially important with a medium vacuum line on which syntheses and purification of inorganic or organometallic materials are routinely carried out. Almost invariably, a rather nasty concoction of debris finds its way through the stopcock orifice into the vacuum chamber over a period of time. The performance of a Teflon stopcock can be seriously impaired under these conditions. Another advantage of glass over Teflon stopcocks with respect to a double manifold vacuum line is that evacuation and refill operations involving small volumes are conducted more easily in one smooth operation. For example, with the Wayda-Dye line mentioned previously (14), two independent Teflon stopcocks must be manipulated to achieve the same result.

Disadvantages in using all glass instead of Teflon stopcocks are considerable, the most important of which is undoubtedly the requirement that they be greased. Grease is often intrinsically troublesome stuff to a practicing synthetic chemist. It channels and thereby allows air to get into the system, contaminates otherwise pure compounds, occasionally reacts with materials and can be easily leached out of stopcocks and joints by solvent vapor. Channelling can be especially problematic if components of a glass stopcock do not exactly match. Grease also traps solvent vapor, which can greatly decrease the vacuum of the system. Another significant problem with all glass vacuum stopcocks is that the components are generally not interchangeable. Often the entire stopcock must be replaced if either the barrel or plug is damaged. Repair of vacuum lines containing glass stopcocks can also be more difficult than those containing Teflon plugs since the line must generally be degreased before repair, particularly if the entire line is to be placed in an annealing oven. Another serious problem is that glass stopcock barrels can be easily warped during repair and/or annealing. Glassblowers are not always as careful as they should be in repairs of this sort, particularly when the repair is made close to the barrel of a stopcock. Anyone who uses high vacuum glass stopcocks should learn how to hand-lap or regrind warped ones for this reason. By mastering this simple technique (which is discussed in some detail on page 163 of reference 1) many trips to the glassblower and considerable money can be saved. A final disadvantage of custom ground glass stopcocks is one of economics: they are considerably more expensive to manufacture than Teflon stopcocks and the cost differential between these two items seems to be steadily increasing. If this trend continues, high vacuum glass stopcocks may be priced out of the market in the not too distant future.

A summary of the approximate prices for the components of the double manifold vacuum line is shown in Table II. A more recent estimate (10/86) from our glassblowing shop (labor: $24/hr)

---

†This is a serious problem with Teflon plugs, especially those which do not have replaceable o-rings, such as ones marketed by Corning glass (Rota-flo stopcocks) and J. Young Ltd. of London.

indicated that it should be possible to construct this double manifold line and accessories (items 1-6) for about $1500. Also shown
is price information on the flexible stainless steel tubing assembly.
The tubing and fittings are available from the Cajon Company, 32550
Old South Miles Road, Solon, Ohio 44139.

Table II. Approximate Cost for Components of the
Vacuum-Inert Gas Line

1.	Molecular Sieve Trap including 3/8 in. Swagelok Union	All items fabricated $108.00 by Glass Tech. Service University of Minnesota
2.	Double Manifold Vac. Line (5 taps)	610.00
3.	Manometers (5) @ 57.50 Cost of the system can be reduced somewhat by incorporating only 2 or 3 manometers. However, the system will be accordingly less flexible.	287.00
4.	Optional Nujol Bubblers (2) @ 59.00	118.00
5.	Cold Trap Assembly including traps	160.00
6.	Glass Connection to Vacuum Pump	30.00

Not included in total cost: Flexible tubing and connections, vacuum
pump, clamps, Dewars and a McLeod gauge.

TOTAL ESTIMATED COST (1984-1985 PRICES) OF LINE: $1,300

7.	Flexible Stainless Steel Tubing Assembly		
	a.	Glass adaptor, 14/35 outer, two 9 mm 1D o-ring seal joints	18.00
	b.	Swagelok Unions, 1/4 x 1/4 B400-6 @ 1.80	7.20
	c.	Flexible SS tubing, 1/4 OD, 24 in. 321-4-X-24 @ 58.50	117.00
	d.	Sleeve Insert for SS tubing 1/4 OD 304-4-XBA @ 2.00	8.00
	e.	Kovar-Glass Seal, 1/4 OD @ 10.00	40.00
	f.	O-ring Seal Joints (4) @ 5.00	20.00
		Est. Total:	$210.00

Specialized Glassware and Techniques

In this final section some of the specialized glassware and tech-

niques we use in conducting bench top organometallic chemistry will
be reviewed.   Included in this section will be a discussion of a low
temperature filtration apparatus, solvent collection unit, solvent
storage flasks and a simple side arm reaction flask which we now use
extensively in our research.

An apparatus which has proven to be particularly useful for
doing liquid ammonia chemistry and the filtering of thermally
unstable organometallics is our low temperature filtration apparatus
depicted in Figure 13.(15)   This design is not commercially available
and has the advantage that the coolant level extends down into the
lower ground-glass joint.   Although we have specified an all glass
stopcock in the design, clearly substitution of a Teflon stopcock
could be advantageous.   Also, the top inner joint may be replaced by
an outer joint which would facilitate the use of septa.   Two different set of dimensions are indicated.   The larger unit is useful
for filtering up to a liter of solution, while the smaller unit is
more convenient for the isolation of solids at low temperatures.

Solvent collection unit and solvent storage flasks which we use
are shown in Figures 14 and 15, respectively.   Improvements of the
solvent collection unit over those commercially available include the
use of vacuum Teflon stopcocks as well as the direct fusion of a
water cooled condenser (or alternatively, a dry ice-acetone condenser
for low boiling solvents) to avoid the use of greased standard taper
joints.   Suppliers often recommend the use of Teflon sleeves or
Teflon coated standard taper joints for greaseless applications,
but we have not found these items to be satisfactorily vacuum tight.
On the top of the condensor is fused another Teflon stopcock which
is connected via a T piece to a mercury bubbler and the vacuum line.
Although we prefer a combination of glass and copper tubing in these
connections, butyl rubber can be used if one is careful to evacuate
and refill several times before admitting inert gas into the distillation unit.   Possible improvements in the design of the solvent
collection unit include the use of an o-ring ball joint for the solvent take off tap, which would prevent stopcock grease contamination,
incorporation of a thermometer well, direct fusion of a Vigreux
column to the solvent collection unit and replacement of the right
angled Teflon stopcock on the reservoir with an Ace Glass "Flickit"
valve (Andrea Wayda is thanked for this suggestion).   The latter
would have to contain Kalrez o-rings to withstand hot solvent vapor,
however.

Freshly distilled solvent can either be stored in the one liter
reservoir of the collection unit or in a solvent storage flask
(Figure 15).   Solvent can be conveniently removed by cannula or
syringe via the right angled stopcock after removal of the Teflon
plug under a countercurrent of nitrogen or argon (a variety of rubber
septa will fit in or over the barrels of these stopcocks, thus per-
mitting transfer of the solvent under positive pressure and fairly
rigorous anaerobic conditions).   In recent years, our group has
utilized even simpler solvent storage flasks which are easily made
by sealing a right angled Teflon stopcock to the top of round bottom,
or Florence flasks (Figure 16).   We have made similar containers from
15 and 40 ml conical shaped graduated centrifuge tubes which are
particularly useful for storage and dispensing of liquid organo-
phosphines, deuterated solvents and other liquids under positive

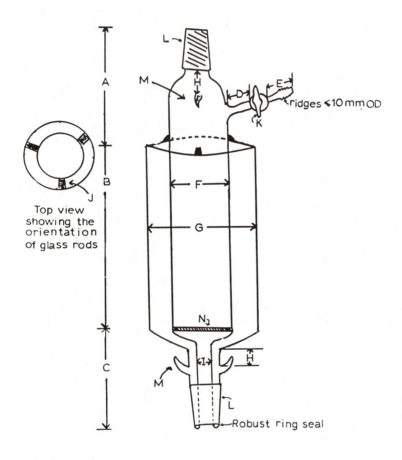

Figure 13.   Low-temperature filtration apparatus. Continued on
next page.

IDENTIFICATION OF
COMPONENTS
AND DIMENSIONS

	LARGE UNIT	SMALL UNIT
A	11 CM	10 CM
B	20 CM	20 CM
C	11 CM	10 CM
D	25 MM	25 CM
E	40 MM	40 MM
F	70 MM OD	35 MM OD
G	120 MM OD	80 MM OD
H	12 MM	12 MM
I	12 MM OD	12 MM OD
J	8 MM ROD	8 MM ROD
K	4 MM †	2 MM †
L	S.T. 34/45	S.T. 24/40
M	STURDY GLASS HOOKS	
N	60 MM FRIT	30 MM FRIT

†NOTE: THE 2 AND 4 MM
OBLIQUE BORE VACUUM
STOPCOCKS MAY BE
REPLACED WITH TEFLON
STOPCOCKS TO GOOD
ADVANTAGE

Figure 13. Continued.

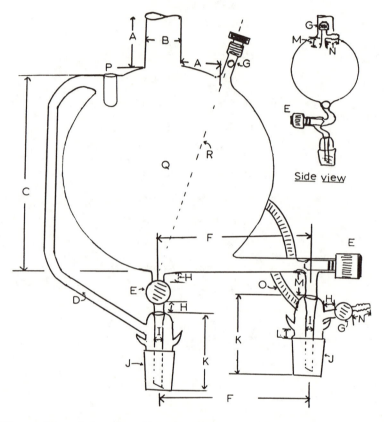

Side view

Figure 14.   Solvent collection unit. Continued on next page.

A	50 MM	J	S.T. 24/40 INNER JOINT
B	25 MM OD	K	75 MM
C	13 CM	L	12 MM
D	15 MM OD	M	25 MM
E	KONTES K826500, 0-8 MM	N	40 MM
F	15 CM	O	8 MM GLASS ROD
G	ACE 8192-03, 0-3 MM	P	THERMOMETER WELL
H	15-25 MM	Q	1000 ML RESERVOIR
I	8 MM OD	R	ORIENTATION OF STOPCOCK F WITH RESPECT TO THE BOTTOM OF THE RESERVOIR TO PERMIT NEARLY COMPLETE REMOVAL OF SOLVENT WITH A NEEDLE

NOTES: 1. EFFICIENT DRY ICE - iPROH AND/OR WATER COOLED CONDENSORS ARE FUSED DIRECTLY ONTO THE RESERVOIR.

2. THE BOTTOM OF THE UNIT IS TO BE CONNECTED BY A S.T. 24/40 JOINT (AS SHOWN) OR PREFERABLY BY DIRECT FUSION TO A 12 INCH VIGREUX OR SIMILAR COLUMN, WHICH IS CONNECTED TO THE SOLVENT POT BY 24/40 JOINTS.

3. ONLY TWO STOPCOCKS ARE SHOWN ON THE SIDE VIEW OF THE UNIT.

Figure 14. Continued.

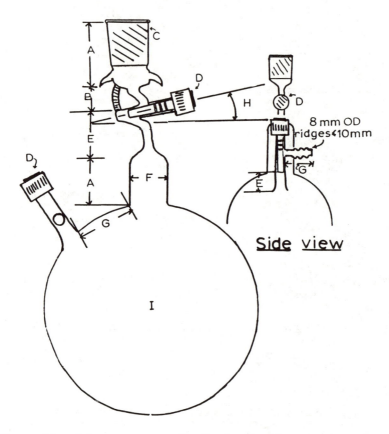

Figure 15.   Solvent storage flask. Continued on next page.

IDENTIFICATION OF

COMPONENTS

AND DIMENSIONS

A 75 MM

B 12–25 MM

C S.T. 24/40 OUTER

D ACE 8192-03

E 25 MM

F 30 MM OD

G 40 MM

H 15–20°

I 500–2000 ML

NOTE: THE SIDE ARM
STOPCOCK IS ORIENTED
SO A NEEDLE WILL REACH
TO THE BOTTOM OF THE
FLASK WHEN SOLVENT IS
REMOVED

Figure 15. Continued.

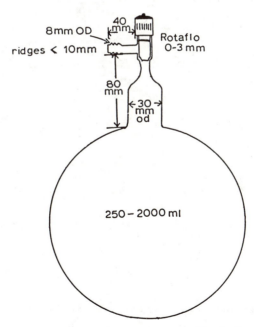

Figure 16.   Inexpensive solvent storage flask from which solvent
or other liquids are dispensed by cannula or syringe.

pressure and strictly anaerobic conditions (Figure 17). By employing these volumetric containers one can avoid the use of syringes
which often contaminate reagents. Also approximate densities of
liquids can be easily determined with this apparatus. Although the
second stopcock may seem superfluous, we have found that the Teflon
stopcocks used for the primary seal occasionally permit slow entry
of air, particularly if the o-ring or Teflon plug is scratched.
Some liquids also attack o-rings. For these reasons the all glass
stopcock (lubricated with Apiezon H or T grease) functions as a
useful "second line of defense" and can be quite important in the
long term storage of highly reactive liquids outside of a drybox.
Of course the best and cheapest method for long term storage of
reactive liquids is to seal them in glass, but this method is not
always practical or convenient.

A particularly useful and inexpensive reaction vessel is shown
in Figure 18. This is simply a side arm, one neck round bottom
flask, equipped with a right angled Teflon stopcock, so liquids can
be added or removed via the barrel of the stopcock as illustrated in
Figure 19. In this fashion one can minimize exposure of an air

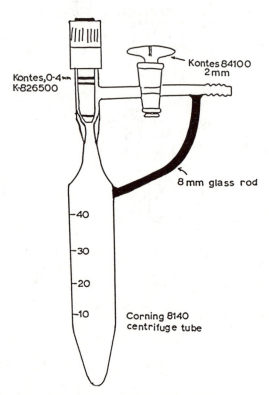

Figure 17. Conical shaped volumetric storage vessel for air
sensitive liquids.

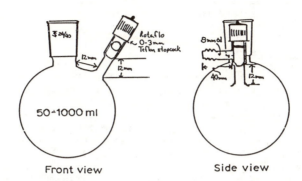

Figure 18.   Reaction vessel incorporating a right angle Teflon stopcock.

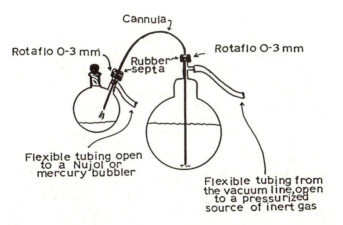

Figure 19.   Illustration showing the transfer of a liquid by cannula from a storage flask into the reaction vessel via the barrel of the right angle Teflon stopcock.

sensitive material to the atmosphere when adding reactants, removing
samples for IR spectra, etc. on the bench. It is much more likely
that aerial oxidation will occur if the sample is removed via the
24/40 joint of the flask than from the stopcock barrel. Exposure
of the reaction mixture to stopcock grease is also minimized in this
fashion.

Concluding Remarks

In this article I have attempted to review the most important
aspects of apparatus and techniques used in our laboratory for the
handling of organometallic compounds. Very little was mentioned
concerning actual procedures in carrying out organometallic reactions or the purification of materials, which are matters of tremendous importance. But other articles in this volume address this
important topic. In closing, however, it is perhaps useful to
stress that success in the challenging, rewarding and rather fast-
paced area of organometallic chemistry requires more than good
equipment, ideas and timing. One must regularly read the current
literature, (experimental sections can be especially helpful), have
a strong work ethic and develop excellent laboratory technique. It
is also important to be persistent, resilient, and have a healthy
measure of dedication and faith that the effort is worthwhile!

Acknowledgements

The National Science Foundation and Petroleum Research Fund, administered by the American Chemical Society, have generously provided
financial assistance for research which directly led to the
development or testing of many of the techniques discussed herein.
I am also very grateful to many dedicated co-workers who have helped
over a period of several years in the design and testing of apparatus. Andrea Wayda is acknowledged for many helpful discussions of
techniques in organometallic chemistry as well as her not so gentle
prodding to finish this article! Also, a sincere thank you to
Professor Du Shriver for his extremely useful book, "The Manipulation of Air Sensitive Compounds", which has served me well for many
years and remains a gold mine of information for all concerned in
this field. Finally, I would like to acknowledge my former advisor,
Professor Alan Davison, who strongly encouraged a creative and
improvisational approach to organometallic and inorganic chemistry.

Literature Cited

1. Shriver, D.F.; Drezdzon, M.A. The Manipulation of Air Sensitive
    Compounds; 2nd Edition; Wiley-Interscience: New York, 1986.
2. Brauer, G., Ed. Handbook of Preparative Inorganic Chemistry; 2nd
    Edition; Academic Press: New York, 1963; Vol. 1 and 2.
3. Dodd R.E.; Robinson, P.L. Experimental Inorganic Chemistry;
    Elsevier: Amsterdam, 1957.

4. Eisch, J.J.; King, R.B. Organometallic Syntheses; Academic Press: New York, 1965, Vol. 1; 1981, Vol. 2.
5. Jolly, W.L., The Synthesis and Characterization of Inorganic Compounds; Prentice-Hall: New York, p. 170.
6. Yamamoto, A. Organotransition Metal Chemistry; Wiley-Interscience: New York, 1986; Ch. 5.
7. Herzog, S; Dehnert, J.; Luhder, K. In Technique of Inorganic Chemistry; Johassen, H.B. Ed.; Interscience-Wiley: New York, 1968; Vol. 7, pp. 119-149.
8. DeLiefde Meijer, H.J.; Janssen, J.J.; Van Der Kerk, G.J.M. Studies in the Organic Chemistry of Vanadium; Institute for Organic Chemistry T.N.O.: Utrecht, Netherlands, 1963; pp. 41-58.
9. Kramer, G.W.; Levy, A.B.; Midland, M.M. In Organic Syntheses Via Boranes, Brown, H.C., Ed.; J. Wiley and Sons: New York, 1975, Ch. 9.
10. Burlicht, J. How to Use Ace No-air Glassware; Ace Glass Inc.: Vineland, New Jersey, 1970; Pamphlet No. 570.
11. Lane, C.F.; Kramer, G.W. Aldrichim. Acta 1977, 10, 11 (Aldrich Chemical Co. Periodical).
12. Gill, G.B.; Whiting, D.A. Aldrichim. Acta 1986, 19, 31.
13. Handbook of Compressed Gases, Compressed Gas Association, Van Nostrand Reinhold: New York, 1981.
14. Wayda, A.L.; Dye, J.L., J. Chem. Ed., 1985, 62, 356
15. Warnock, G.F.P.; Ellis, J.E., J. Am. Chem. Soc., 1984, 106, 5016 and references cited therein.
16. Kelsey, B.A.; Ellis, J.E.; J. Am. Chem. Soc., 1986, 108, 1344; Chi, K.M.; Frerichs, S.R.; Stein, B.K.; Blackburn, D.W.; Ellis, J.E.; submitted for publication.

RECEIVED July 31, 1987

Chapter 3: Application 1

Preparation and Isolation of Crystalline Samples Using Low-Temperature Solution Techniques

Malcolm H. Chisholm and David L. Clark

Department of Chemistry, Indiana University, Bloomington, IN 47405

Attached is a schematic drawing (Fig. 1) of a low-temperature reaction vessel for the high yield preparation and isolation of thermally unstable compounds. The dimensions of a typical reaction vessel are summarized in a). The Kontes valve may either be sealed directly to the glass, or connected via Tygon tubing depending on the air sensitivity of the materials used. The reactant solution is placed on the frit, and the Kontes valve closed under nitrogen. The closed end of the vessel is then immersed in liquid nitrogen just below the frit as shown in b). This provides the vacuum to pull the solution through the frit and into the closed end of the vessel where it is frozen at -196°C. The volatile reactant is then condensed into the vessel, and the vessel sealed with a torch as indicated in b). The vessel may then be warmed to the desired reaction temperature and ultimately cooled to -78°C in a Dewar of dry ice as shown in c). This procedure obviously mandates the use of a solvent such as toluene to avoid freezing at this temperature. Ideally, crystals will grow at -78°C as indicated in c). When crystals are present, the vessel is inverted, and placed in liquid nitrogen just below the constriction to pull the filtrate solution into the receiving bulb. This procedure will deposit crystals on the frit. The vessel is then sealed at the constriction leaving the filtrate in the bulb and crystals in a fritted ampule as shown in d). Caution: care should be used to get a uniform and tempered seal in step b) to avoid cracking when immersing in liquid nitrogen.

RECEIVED August 13, 1987

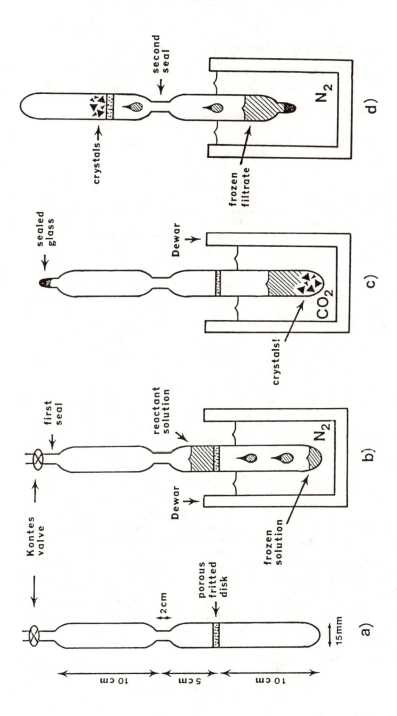

Figure 1. Schematic drawing of a low-temperature reaction vessel for the high yield preparation and isolation of thermally unstable compounds.

Chapter 3: Application 2

Recrystallization Apparatus for Air-Sensitive Compounds

Alan E. Friedman and Peter C. Ford

Department of Chemistry, University of California, Santa Barbara, CA 93106

Described here is a crystallization apparatus that uses vapor transfer as the method of slowly exchanging the solvents. This technique allows small quantities of air and moisture sensitive organometallic complexes to be crystallized easily under an inert atmosphere to give X-ray diffraction quality crystals.

A Schlenk tube (Figure 1) is modified so that a deep dish sample vial can be easily placed inside in a small glass cup to provide stability. For convenience a ring-like glass handle is placed on the sample vial to assist in its removal once crystals have formed. The 24/40 stopper of the Schlenk tube is also altered with a hook placed on the inside to hoist the sample vial. Lastly, a syringe port is set on the side of the Schlenk tube to provide access to both the outside and inside of the sample vial.

A recrystallization is accomplished by placing a sample of the solid material into the sample vial which is then placed into the Schlenk tube. The Schlenk tube is evacuated and/or flushed with the appropriate oxygen free gas by several pump/fill cycles. A minimum of degassed solvent in which the sample is soluble is then added to the sample vial via syringe techniques through the septum covering the syringe port. A second degassed solvent, in which the sample is known to be insoluble, is then placed in the outside part of the Schlenk tube so that gaseous diffusion from the outside to the sample containing vial can occur. After a few days crystals will begin to form. For materials that are thermally sensitive the apparatus may be placed in a refrigerator or freezer during this time. Use of this apparatus may be modified for assembly within a dry box.

The advantage of this technique over similar methods is the compactness of the apparatus allowing it to be oven dried and its convenience for use with vacuum lines and syringe techniques. Another advantage is the relative convenience by which crystals can be removed from the vial. One main limitation is the small size of the inner vial which may overflow if the "outer" solvent is the more volatile of the two.

RECEIVED July 31, 1987

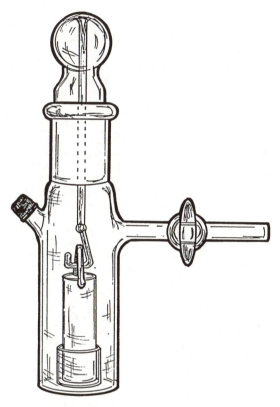

Figure 1. Crystallization Apparatus: The hollow 24/40 stopper is fitted with a hook that extends 5 cm below to the ring of the 1 cm sample vial. The syringe port is made from 7 mm tubing while a 2 mm hollow plug, high vacuum stopcock is used on the sidearm. The diameter of the main body is 3.5 cm while the overall length is 20 cm.

Chapter 3: Application 3

Inert Atmosphere Apparatus
for UV Photochemical Reactions

William C. Trogler

Department of Chemistry, D-006, University of California at San Diego, La Jolla, CA 92093

Small (1-5 g) scale UV photolysis of air sensitive compounds can be performed in quartz Schlenk tubes, or in conventional Schlenkware with the use of a UV transparent quartz stopper. The latter apparatus is easily adapted to low temperature irradiations. Large scale (10-50 g) UV photochemical reactions use quartz immersion well reactors. Medium pressure Hg-arc lamps are the preferred radiation sources for synthetic applications.

This report summarizes conventional methods for UV irradiation of air sensitive organometallic compounds at ambient or subambient temperatures. Of the irradiation sources available (1) the medium pressure Hanovia 450 W arc lamp systems (2) are of moderate price, reliable, and versatile in our experience. **Caution: Powerful arc lamps can cause eye damage or blindness within seconds and UV protective goggles (available from most scientific supply houses) must be worn. Never look directly at the radiation source. For safety of other workers lamps should be enclosed in a vented box with baffles.** If Pyrex transmits enough UV radiation for an efficient reaction, as for photochemical reactions of metal-metal bonded complexes (3), then conventional Schlenkware can be used for photolysis and no special glassware is needed. Since a 2 mm thick wall of Pyrex transmits only 10% of the UV light at 300 nm, UV transparent quartz reaction vessels are often needed for photoreactions of mononuclear organometallic complexes.

Quartz Schlenk tubes are inexpensive if made from standard quartz tubing closed at one end, and attached to a graded seal. The desired stopcock and joint are attached to the Pyrex end of the seal. Stock tubing and seals with 1 cm I.D. to 3-1/2 inch I.D. are available commercially (4). Although the curved surface of a quartz tube reflects much light, the apparatus is effective for moderate scale reactions (<5 g) of high quantum efficiency. For large scale reactions (10-50 g), or for reactions of low quantum efficiency, an immersion well system works best in our experience (5). In this arrangement a cylindrical arc lamp, contained in a water-jacketed quartz tube (2), is immersed in a Pyrex reaction flask (Figure 1).

0097-6156/87/0357-0070$06.00/0
© 1987 American Chemical Society

The minimum volume assembly shown in Figure 1 consists of a 60/50 outer joint attached to 23 cm of 50 mm I.D. Pyrex tubing. A 14/20 outer joint attached to the upper sidearm is septum capped for the use of cannula techniques, as described elsewhere in this volume. It is most convenient to purge the assembled apparatus with N_2 before introducing the reaction solution by cannula. The glass inlet tube at the bottom of the flask permits the introduction of gas, before and during photolysis (a syringe needle in the septum serves as a vent), or the removal of solvent by applying gas pressure through the 14/20 sidearm. A medium frit-Teflon value sidearm at the bottom of the flask permits filtration of reaction solutions directly into a Schlenk flask for further work up. Solid residues can also be dissolved and filtered through the sidearm for recrystallization. Larger volume reaction flasks (250-1000 mL) are constructed with large diameter glass tubing, and other designs are available (2). We have used the apparatus (5) of Figure 1 to prepare 40 g lots of substituted metal carbonyls with about 15 h of photolysis. When using the immersion well system it is important to stir the solution with a large magnetic stir bar so the solution in the irradiation dead volume at the bottom of the flask circulates around the sides of the lamp well. In some reactions a nontransparent film of decomposition products forms on the outer surface of the quartz well. For moderately air sensitive compounds, a vigorous purge of N_2 through the upper side arm can be used to remove the well for periodic cleaning during reaction. For photoreactions that generate gases, such as CO, the evolved gas is vented through a tubing adapter on the 14/20 sidearm to a mineral oil bubbler. As a rough method of quantification a Tygon tube may be attached to the bubbler and placed in the mouth of an inverted water-filled graduated cylinder, which is placed in a partially filled beaker of water (5). The evolved gas (if of low solubility in water, as for CO and H_2) displaces water in the graduated cylinder and the extent of reaction can be estimated with reasonable accuracy. It is crucial to carefully monitor the progress of UV photoreactions because the products usually also absorb UV light and decompose if conversions are pushed too far.

One advantage of photochemistry is the ability to produce a highly reactive intermediate at low temperature so a clean reaction occurs. While a UV transparent alcohol can be used as a recirculating coolant for the quartz well system described above, the outer Pyrex reaction flask must also be cooled; temperatures below $-30^{\circ}C$ are not obtained easily. An alternative approach of immersing a quartz reaction flask into a dewar with quartz windows, or one constructed with a quartz body, suffers because of scattering and absorption by most dry ice/solvent slushes. One solution to the problem (Figure 2) requires a collimated source of UV light from a 200-500 W Hg-Xe or Hg short arc lamp source with fused silica optics (6). An additional focusing lens (~50 mm) is recommended to produce a tighter beam shape for passage through the transparent stopper described below. A Schlenk flask is modified with a 29/26 outer joint used in place of the usual 24/40 outer joint at the upper neck of the flask, and a small septum capped sidearm is added to provide a cannula inlet for synthetic manipulations. A 1-1/4 inch diameter quartz disk (4), which is 1/8 inch thick, is fixed to a 29/26 joint sawed just above where the glass tubing connects to the joint. This

Figure 1. An immersion well apparatus consisting of a cylindrical 450 Watt Hanovia medium pressure arc lamp in a water-cooled quartz well. The quartz well is shown surrounded by the Pyrex reaction flask.

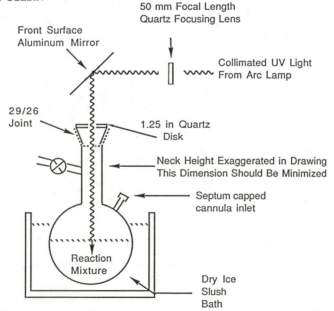

Figure 2. Schematic drawing of apparatus for low temperature UV irradiation of air sensitive solutions.

acts as a UV transparent stopper for the Schlenk flask. The collimated beam from the arc lamp (after filtering through a 10 cm path distilled water filter to remove IR radiation) is focused to a smaller beam shape and bounced off a front surface aluminum mirror (7) held at 45° from vertical. This directs the light beam through the transparent stopper into the Schlenk flask. The mirror-stopper distance and neck height of the Schlenk flask should be minimized. A low-form dewar filled with the desired slush bath surrounds the Schlenk flask and does not interfere with the irradiation. This latter approach also provides a convenient method for room temperature photoreactions of air sensitive solutions on a modest (1-3 g) scale.

Literature Cited

1. Summaries of experimental apparatus may be found in Chapter 7: Calvert, J. G.; Pitts, J. N., Jr. Photochemistry; John Wiley & Sons, Inc.: New York, 1966, and in Rabek, J. F. Experimental Methods in Photochemistry and Photophysics. Part II; John Wiley & Sons: Chichester, 1982.
2. Available from Ace Glass Inc., P.O. Box 688, 1430 Northwest Blvd., Vineland, New Jersey 08360.
3. Schmidt, S. P.; Trogler, W. C.; Basolo, F. Inorg. Syntheses 1985, 23, 41.
4. Available from Quartz Scientific Inc., 819 East Street, Fairport Harbor, Ohio 44077.
5. Therien, M. J.; Trogler, W. C. Inorg. Syntheses, in press.
6. Available from Kratos Analytical, 170 Williams Drive, Ramsey, New Jersey 07446 or from Oriel Corporation, 15 Market Street, Stamford, Connecticut 06902.
7. Inexpensive mirrors and other optical equipment are available from Edmund Scientific Company, 101 East Gloucester Pike, Barrington, New Jersey 08007.

RECEIVED July 31, 1987

Chapter 3: Application 4

Distillation Flask for Use with a Heating Mantle

Hugh Felkin

Institut de Chimie des Substances Naturelles, Centre National de la
Recherche Scientifique, 91190 Gif-sur-Yvette, France

Although electric heating mantles are very convenient sources of
heat, they are generally used for refluxing rather than for
distillation, since they have the disadvantage of overheating the
walls of the lower half of the flask as soon as these have gone to
dryness. This can be hazardous if for example the substances being
distilled contain traces of peroxides, or a drying agent such as
lithium aluminum hydride.
  The modified flask shown in the Figure, which is easily made
from, e.g., a 2000 ml flask and a 250 ml flask, allows the
distillation, with a heating mantle, of all but about 120 ml of
the contents without overheating of the walls. It is particularly
useful, in conjunction with a suitable reservoir and reflux
condenser, for the continuous refluxing and distillation of dry,
degassed, solvents. A further advantage of using a small heating
mantle with a large flask is that the apparatus requires no
supervision on start-up, since, whatever the setting of the
heating mantle control, it is generally impossible for the heat
input to exceed the cooling capacity of the condenser.

RECEIVED July 31, 1987

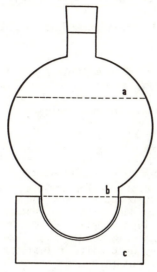

Figure 1.   Modified distillation flask.   (a) Maximum level;
(b), minimum level; (c), electric heating mantle.

Greaseless Stillheads for Sensitive Solvent Purification

Andrea L. Wayda and Patricia A. Bianconi

AT&T Bell Laboratories, Murray Hill, NJ 07974

A modified stillhead (based on a commercially available apparatus) is described which allows the facile grease-free distillation of small and large quantities of solvents under anaerobic conditions.

Work with very air and moisture sensitive organometallic compounds requires solvents which have been rigorously dried and degassed. While methods vary according to solvent type, such purifications are routinely effected by distillation under an inert atmosphere from a water and oxygen scavenging reagent such as sodium benzophenone ketyl. In those cases where the solvent is reactive toward this reagent, purification is accomplished by distillation from a drying agent followed by degassing by purging the solvent with an inert atmosphere stream or by successive freezing and pumping of the solvent. Since these manipulations can be tedious, time-consuming and occasionally ineffective with improperly designed glassware, we describe herein a solvent stillhead design which utilizes easily modified commercially available glassware and which provides for rapid and efficient distillation and handling of a large or small quantity of solvent.

The Kontes anaerobic stillhead (1) is available in two convenient sizes and incorporates two extremely useful features: 1) a compact one-piece design which eliminates multiple connections at which leaks may develop and 2) the incorporation of Teflon valves throughout to provide leak-free, grease-free seals. It is also versatile since small quantities of solvent may be purified by distillation and withdrawn by syringe after collection in the solvent reservoir or larger quantities may be distilled or drained into an attached receiver flask. However, in our work with the stillhead, we have found that its performance could be substantially improved by incorporating the following design modifications (Figure 1): 1.) attachment of a 4 mm Teflon valve to the inert gas inlet (A) which allows for isolation of the still from the handling manifold, 2.) replacement of the screw-cap septum outlet with an Ace "Flickit" valve (2) (a quick-open Teflon valve with an unobstructed 5 mm pathway when open (B)) which allows repeated syringe punctures or changes of the septum without breaching the integrity of the

0097–6156/87/0357–0076$06.00/0
© 1987 American Chemical Society

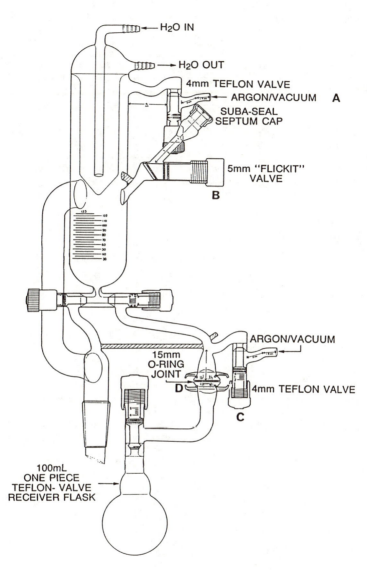

Figure 1. Modified 125 mL capacity Kontes anaerobic stillhead.
Note the design modifications labeled A, B, C, and D and described
in the text. (Illustration kindly provided by Crown Glass Company.)

still and 3.) modification of the solvent take-off arm by attachment of a 4 mm Teflon valve (C) and 15 mm O-ring joint (D) thereby providing for independent evacuation and backfill of the receiver flask and for greaseless transfer of the solvent, respectively.

By incorporating these features in our modified stillheads, we can routinely collect small (10-25 mL) or large (1 L) quantities of solvents with a minimum of manipulation. In addition, the stills are easy to maintain (fresh solvent is added under strong argon purge into the spare port of a 2-liter reservoir flask) and retain their integrity indefinitely (as judged by the retention of the strong purple color of active sodium benzophenone ketyl). Note, however, that care must be taken to match the exposed O-rings used in the stillhead with known solvent compatibilities. Viton is the general elastomer of choice for most solvents. However, **O-rings in direct vapor contact with polar solvents such as THF or ether must be replaced with Kalrez (3)** if an inert atmosphere is to be retained in the still.

REFERENCES

1) Kontes, Spruce Street, P.O. Box 729, Vineland, New Jersey 08360.
   K-547600-0125, 125 mL distilling head. ($166.60)
   K-547600-0500, 500 mL distilling head. ($220.00)

2) Ace Glass Inc., P.O. Box 688, 1430 Northwest Boulevard, Vineland, New Jersey 08360. 8200-05, "Flickit" valve, 5 mm size, $62.15.

3) In practice, only the seat O-ring of the "Flickit" valve is replaced with Kalrez. Other valves use Teflon plugs without a seat O-ring and are backed by Viton O-rings. Since Kalrez does not have the resiliency of Viton, the "Flickit" valve O-ring seat should be inspected periodically to ensure that its shape (and the seal) have been retained.

RECEIVED July 31, 1987

Chapter 4

Vacuum Line Techniques for Handling Air-Sensitive Organometallic Compounds

Barbara J. Burger and John E. Bercaw

Division of Chemistry and Chemical Engineering, California Institute of Technology, Pasadena, CA 91125

Vacuum line techniques for handling air sensitive oganometallic compounds are described. The vacuum line is discussed in six sections: (i) the pumps and main traps, (ii) the main manifold with pressure gauges, (iii) work stations, (iv) the cryogenic traps and Toepler pump, (v) the inert gas purifiers and inlet system, and (vi) the specialty gas inlet system. Use of a swivel frit assembly in the synthesis of air sensitive compounds is described along procedures for the addition of gases, volatile and nonvolatile liquids and solutions to reaction mixtures. High temperature and pressure reactions are discussed in terms of the use of sealed NMR tubes and heavy walled reaction vessels. The procedure for molecular weight determination using the Signer method is also included.

Our research group carries out preparative chemistry in fume hoods, on bench tops, in high pressure (Parr) reactors, in glove boxes, in Schlenk tubes, and with cannulas and septa, just as other groups concerned with the synthesis and characterization of air-sensitive organo-transition metal compounds. In this article we attempt to describe the less common methods, procedures and the equipment, i.e. those associated with a vacuum line, which have evolved over the past fifteen years in the Bercaw group. Many of the techniques are taken from the literature and have been (hopefully) improved upon or modified for working with small amounts (50 mg – 35 g) of an organotransition metal compound. Others have evolved out of necessity, and, as far as we are aware, are not described elsewhere.

0097–6156/87/0357–0079$06.00/0

The Vacuum Line

Most of the reaction chemistry that is done in our
laboratory involves the use of vacuum lines, such as the
one shown below in Figure 1. The use of a vacuum line
offers two major advantages over more standard Schlenk
techniques: hoses and rubber septa, which are more
permeable to air and moisture, are completely excluded
and quantitative manipulation of gases and volatile
liquids can be performed conveniently. Although the
construction of a vacuum line as complex as that shown
involves a relatively large investment of time and money
(ca. $15,000 in parts and labor), it becomes a permanent
piece of equipment which is shared by two or three group
members and used extensively on a regular basis.
Moreover, it is often faster to assemble a complicated
apparatus on the vacuum line, rather than at a Schlenk
line, since many of the required components are
permanently in place.
     The vacuum line consists of six main sections: (i)
the pumps and main traps, (ii) the main manifold with
pressure gauges, (iii) the work stations, (iv) the
cryogenic traps and Toepler pump, (v) the inert gas
purifiers and inlet system, and (vi) the specialty gas
inlet system.
     Each vacuum line is equipped with two pumps which
operate in series. A mercury diffusion pump, together
with a mechanical fore pump, provide a residual pressure
of <10^{-4} torr, which is adequate for the syntheses that
we undertake. Two traps cooled by large dewars filled
with liquid nitrogen are placed, one before and one after
the diffusion pump to protect the pumps from condensibles
and the mechanical pump from mercury. The mechanical
pump is exhausted to the hood. The residual pressure in
the manifold is measured by the use of either a rotating
McLeod gauge or a thermocouple gauge. The McLeod gauge
provides the more accurate measurement of the residual
pressure in the system, especially at pressures <10^{-3}
torr(1). Disadvantages with this gauge are that it needs
to be regreased and cleaned regularly, and since it is
slow, it is not as convenient when one is trying to
locate a leak in the line. Although they periodically
need to be recalibrated, the thermocouple gauges are
relatively inexpensive and provide a continuous
indication of residual pressures >10^{-3} torr.
     The main manifold of the vacuum line is kept
evacuated and isolated from the work stations on the line
by stopcocks which are lightly greased with Apiezon N
Grease. There are two types of work stations: a
standard Schlenk line section, consisting of vacuum and
inert gas manifolds connected by three way stopcocks to
standard taper 24/40 inner joints, and a high vacuum work
station for working with smaller quantities of material
or for accessing the cryogenic traps and Toepler pump
section. The two manifolds of the Schlenk line work

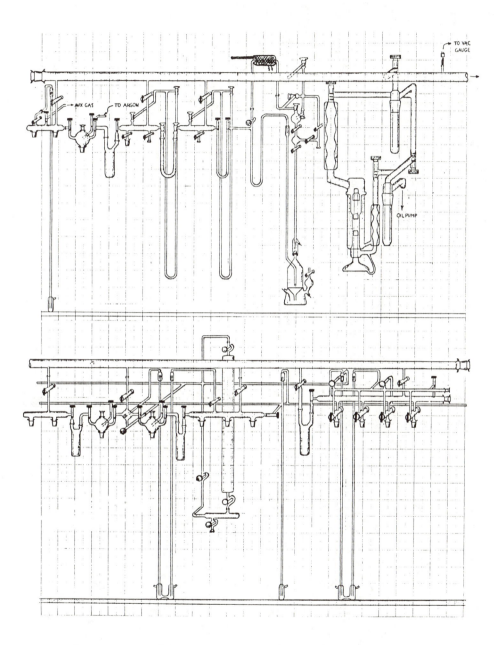

Figure 1.   Diagram of a Bercaw group vacuum line
(courtesy of Dr. Dean M. Roddick).

station are equipped with mercury bubbler/manometers,
which serve two general purposes: (_i_) to provide a rough
measure(2) of the amount of gas that is admitted to a
manifold and (_ii_) to provide an exit when the pressure in
the manifold is excessive. Gas is exited from the
bubbler into tygon tubing affixed to a hood exhaust. A
separate mechanical pump for the vacuum manifold of the
Schlenk line work station is used for several of our
lines, allowing two people to work simultaneously at the
line without interfering with each other. The vacuum
manifold is also connected (_via_ a large stopcock) to the
main manifold of the high vacuum line, so that access to
high vacuum is still possible. Similarly, the inert gas
manifold is connected (via a stopcock and manifold) to
the specialty gas section of the main line. Large
capacity, removable traps, affixed to the end of the
vacuum manifold by means of O-ring joints, are used to
collect the large volumes of solvent which are removed in
vacuo from reaction mixtures. The standard taper 24/40
outlets may be fitted with butyl rubber tubing with an
outer joint. Thus, the Schlenk line may be operated with
its own rough pump in a conventional mode, or, when
connected to the main vacuum and specialty gas system,
more like the work station described below.

The high vacuum work station is the heart of the
line. It consists of two main 24/40 standard taper inner
joints connected via teflon needle valves(3) to (_i_) the
high vacuum manifold, (_ii_) the inert gas manifold, (_iii_)
the specialty gas manifold, (_iv_) each other, (_v_)
manifolds leading to large volume, removable traps, (_vi_)
outlets on these smaller manifolds for admitting gases
from bulbs or lecture bottles or solvents, reagents, etc.
from flasks or calibrated volumes, and (_vii_) at the right
to the cryogenic traps and Toepler pump system. The
large number of connections provided to these ports
presents a design problem, which we have solved by
incorporating 100 ml flasks as connectors. Glass strain
at the connections is thus relieved by spreading out the
area of the glass seals. Moreover, the flasks serve as
splash guards and provide easy access for cleaning
through their stoppered 14/20 standard taper outer joints
at the top. The entire work station, as well as most of
the other manifolds, Schlenk line, bubblers, _etc._ are
interconnected and connected to the line via O-ring
joints for easy removal when cleaning.

The cryogenic trap section of the line consists of
one more large capacity, removable trap, and three U
traps leading to the inlet of the Toepler pump. These
traps can be used to separate volatile materials on the
basis of differing vapor pressures, as is performed
routinely for boranes, phosphines, _etc._ on lines similar
in design to this section of the vacuum system.
Normally, we use these traps to assure complete removal
of solvent and other liquid nitrogen condensable
compounds from noncondensables such as CO, H_2, CH_4 and

N_2, which are quantitatively collected by the Toepler pump. Needle valves are used to allow only small pressures of gas to flow into the Toepler pump, in order to prevent damage from a "mercury hammer"(4). The pressure is indicated by manometers at one or two points along the manifolds connecting the traps. These manometers are also useful in measuring out known amounts of reagents or gases into calibrated bulbs, which may be attached to outlets along the manifolds. The Toepler section consists of the pump, several precalibrated volumes separated by 2-way, oblique bore stopcocks, an outlet for attaching a gas infrared cell, mass spectrometry sample bulbs, etc., a stopcock to the main high vacuum manifold, and a recirculating manifold(5), a section of which is packed with CuO that may be heated to 320°C (vida infra). The CuO section of the tube is wrapped, along with a glass tube to hold a 360°C thermometer, with heating tape which is connected to a Variac transformer, then encased in thermal insulating material. The inlet tube of the Toepler pump and the first, tubular calibrated volume of the pump are positioned parallel to each other so that a section of meter stick may be taped into place. Thus, when the Hg level is allowed to rise to a mark above the top valve assembly, the inlet tube functions as the Hg column of a manometer for reading the pressure in the calibrated volumes.(6) The Toepler pump cycle is automated by using the mercury as a switch alternately connecting the common lead to the top lead (which activates the solenoid valve to open the lower chamber to a mechanical pump) and the lower lead (which activates the solenoid valve to open the lower chamber to a slow, filtered air lead. The Toepler pump control box is assembled at Caltech and is comprised of a 6.3 V transformer (e.g. Triad F 14X), a relay (e.g. PD KRP11AG), a thermal delay relay (e.g. 6N02), a 2 ohm 10 W resistor, five female AC plugs, chassis mount (for the mechanical pumps, diffusion pump heater, CuO tube heater, etc.), five tineon lamps, 125 V, 0.3 W, five SPST 10 A power switches, a fuse holder, an amphenol 5 pin hex chassis mount female connector (to receive male connector with wires from alligator clips attached to leads on Toepler pump), a 6 foot power cord, 3 wire, number 16 or larger, and two 4 pin terminal strips all contained in an aluminum chassis with cover plate (3X5X10).(7)

The two working inert gases, dinitrogen and argon, are taken from 1A cylinders through regulators, copper tubing and metal to glass seals and passed through purifiers before they enter the vacuum line. Two large capacity columns (Figure 2), one filled with MnO supported on vermiculite(8), the other with 4Å molecular sieves, remove dioxygen and water, respectively. Periodically (approximately annually) the oxidized MnO (of empirical formula Mn_3O_4) needs to be reduced. Reduction is accomplished conveniently by passing

Figure 2.   MnO/vermiculite and molecular sieve columns
            for argon (left) and dinitrogen (right)
            purification.

dihydrogen (hood exhaust) through the column while
heating to 300-350°C. The column filled with molecular
sieves is normally regenerated at the same time by
passing dinitrogen at approximately 250°C. The columns
may be heated using commercially available heating tape,
or by wrapping the inner column with nichrome wire.
Strips of asbestos tape on the sides, one next to the
glass column, the other over the nichrome wire are
essential to hold the wire in place. The asbestos tape
is wetted to adhere it to the column and wire. The
resistance of the wire is chosen to maximize the power at
120 V and 5 A, the maximum settings for a standard Variac
transformer. The glass outer jacket of the column
reduces the cooling effects of drafts. Wrapping the
outer jacket with aluminum foil(9) during this process
further increases the heating efficiency. The normal
working inert gas, Ar, is transferred through glass
tubing to the Ar manifold of the two working stations. A
mercury bubbler is connected to this manifold and the Hg
level and regulator set to maintain ca. +50 torr Ar
pressure without continual bubbling. The dinitrogen is
usually transferred through glass tubing to the specialty
gas manifold described below. N_2 is the inert gas of
choice when working with an apparatus cooled near liquid
nitrogen temperature(10).

A smaller, all-in-one column, filled with
MnO/vermiculite (lower half) and molecular sieves (upper
half) is used for purifying specialty gases such as
dihydrogen, dideuterium, and carbon monoxide(11). This
apparatus is attached to the vacuum line near the inert
gas manifold, as shown in Figure 1. It is a good working
practice to leave the column filled with the last
specialty gas used, with a label indicating the column
contents. Replacing the gas by evacuating the column
packed with vermiculite and molecular sieves is quite
time consuming.

Solvent Pots and Their Use

When using a vacuum line, solvents are introduced by
vacuum transferring from a solvent pot into the reaction
flask. To prepare a solvent pot, the solvent is
rigorously dried by refluxing over an appropriate drying
agent in a solvent still. Drying agents such as
"titanocene"(12) or sodium and benzophenone (for
preparation of Na/benzophenone-ketyl) and a stir bar are
added to a large round bottomed flask (~ 500 ml). The
flask is then attached to a 180° needle valve adapter,
which consists of a 24/40 standard taper inner joint,
straight needle valve and 24/40 standard taper outer
joint as shown in Figure 3. With exclusion of air,
solvent is transferred from the solvent still to the
solvent pot (transfer is facilitated greatly by
evacuating the solvent pot first(13)). The solvent pot
is then degassed by attaching to the high vacuum line

Figure 3.   High vacuum line work station with solvent
            pot (left) and swivel frit assembly
            (right).

and, while stirring at room temperature, pumping on the
solvent through a -196°C trap for a couple of minutes.
To transfer solvent from the pot into a reaction flask,
the entire assembly interconnecting the two is completely
evacuated. Cooling the reaction flask with either a dry
ice/ethanol bath (-78°C) or with liquid nitrogen
(-196°C) and slowly opening the needle valve of the
solvent pot while stirring magnetically causes solvent to
transfer into the flask. The rate of transfer will
depend on: (i) the vapor pressure of the solvent, (ii)
the residual pressure in the manifold and the temperature
of the solvent pot. Keeping a room temperature(14) water
bath under the solvent pot will facilitate the transfer.

Swivel Frit Assemblies and Their Use

Preparative synthesis on a vacuum line is carried out
primarily using a swivel frit assembly. Although they
may be constructed in virtually any size, a typical frit
assembly has 14/20 standard taper joints and can
accommodate up to approximately 10-15 grams of solid. A
diagram of an apparatus of moderate size is shown below
in Figure 4. A quick opening teflon valve or a stopcock
is centered in the pressure-equalizing side arm, to
isolate one side of the frit from the other. An adapter
consisting of a 14/20 standard taper inner joint, right
angle teflon needle valve and 24/40 standard taper outer
joint, is used to attach the assembly to a port of the
working section of the vacuum line. In a typical
experiment, one of the round bottomed flasks is charged
with solid reagents (these are loaded in the inert
atmosphere glove box if they are air sensitive), and a
magnetic stir bar is placed in both flasks. The joints
of the frit and the needle valve are greased lightly with
Apiezon H grease(15) and are secured with springs or
rubber bands. With the teflon needle valve tightly
closed, the assembly is removed from the glove box and
attached to the vacuum line at one of the ports of a
working station via 24/40 standard taper joint, lightly
greased with M or H grease. The system is evacuated,
first up to the teflon needle valve of the adapter, then
including the entire frit assembly. An appropriate
solvent is then transferred into the reaction flask
following the procedure described above.

Adding Reagents to the Reaction Flask of the Swivel Frit
Assembly

Condensable Gases. To add a desired quantity of a
condensable gas, a a gas bulb of known volume is normally
used. The working section, gas bulb and connection to
the gas supply, manometer(16) and frit assembly are all
evacuated (see Figure 1). If the frit assembly contains
solvents or volatile materials, the lower flask is cooled
to -78°C or -196°C before evacuation. The frit assembly

is then isolated from the ensuing procedures by closing
the needle valve of the adapter. Gas is admitted to the
desired pressure calculated using the simplified
expression of footnote 6. The stopcock of the calibrated
bulb is closed and the gas remaining in the manometer and
the rest of the working section is removed by evacuation
by condensing it back into its storage bulb or evacuating
the system through a removable trap(17). The working
section is then isolated from the vacuum manifold by
closing the stopcock to the main manifold, and the gas in
the bulb is condensed into the reaction flask by cooling
to -196°C. If the gaseous reagent has been purified by
freeze-pump-thaw cycles prior to use and is stored in a
-196°C trap, the desired quantity of gas can be admitted
into the manifold and calibrated gas bulb by removing the
liquid nitrogen coolant and slowly warming to room
temperature until the desired quantity has been admitted
to the manifold. The need for a calibrated gas volume
can be eliminated by determining the volume of the
manifold and using it as the calibrated gas volume,
although a manometer correction must then be applied to
account for the increase in volume with increase in
pressure.

Volatile Liquids. Volatile liquids can be added to a
reaction flask of a frit assembly by either of two
methods, depending on the quantity of liquid being added
and its vapor pressure at room temperature. The first
method is analogous to that used for gaseous reagents,
and thus is suited to small amounts of very volatile
compounds. The liquid to be added is rigorously dried,
degassed and stored (in vacuo) in a solvent container
such as a round bottomed flask with adapter consisting of
a standard taper inner joint, 180° teflon needle valve
and 24/40 standard taper outer joint. The liquid is then
admitted into a calibrated gas bulb and condensed into
the frit assembly as described above for a gaseous
reagent. For larger amounts of liquids, a graduated
volume(18) attached to an adapter consisting of a
standard taper inner joint, 180° teflon needle valve and
24/40 standard taper outer joint is useful (Figure 5).
The desired amount of liquid is vacuum transferred into
the graduated volume. Once this is accomplished and the
manifold is evacuated (either by condensing any residual
vapor back into the solvent container or into a -196°C
trap), the liquid is then vacuum transferred into the
reaction flask of the frit assembly. For storing liquids
which are low boiling (just above room temperature) or
which are not compatible with Apiezon greases, a solvent
container such as that shown in Figure 6 is convenient to
use. This container is similar in design to that
described later in the section on thick-walled reaction
vessels except that it is constructed from standard wall
thickness glass tubing. The liquid to be stored is first

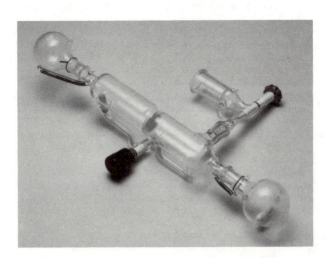

Figure 4.   Swivel frit assembly.

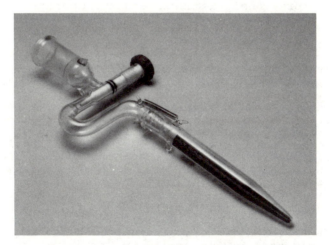

Figure 5.   Graduated volume for addition of volatile
            liquids to reaction flasks.

purified by the appropriate means and then vacuum transferred into the container for storage under vacuum.

Solutions and Non-Volatile Liquids. In many instances, it is desirable to add a solution (such as a commercially available solution of a Grignard or alkyl-lithium reagent) or a non-volatile liquid such as dimethylphenylphospine to a reaction flask. This addition to the reaction flask of the swivel frit assembly can be accomplished in the following manner: under an argon counterflow, the teflon plug from the right angle teflon needle valve of the adapter(19) is removed. A syringe attached to a length of polyethylene tubing (approximately six inches in length) is inserted into the needle valve and degassed by drawing argon through the syringe several times. Filled with argon, the syringe is disconnected from the tubing. The syringe is inserted into the reagent container (usually stored under an inert atmosphere and capped with a sure-seal top or a septum) and the desired volume drawn out. The syringe is then reattached to the tubing and the solution syringed into the reaction mixture. The teflon plug is then fit back into the adapter, leaving the reaction vessel open to the mercury bubbler/manometer. This same technique can be used for non-volatile liquids , which have been previously degassed and are stored under an inert atmosphere.

Filtering Using a Swivel Frit

When working up a reaction mixture, an unwanted solid (salt, etc.) often needs to be removed from a solution of the desired product. Alternatively, the product of a reaction might be insoluble directly from the reaction mixture. In either case, a filtration is required. There are two general ways in which to do this; both rely on differences in pressures to drive the filtering process. The first method is especially convenient when filtering a hot solution. The entire frit assembly is filled with argon. Prior to filtration, the valve of the side arm is closed and the assembly is inverted. A slight vacuum is pulled on the bottom flask causing the solution to pass through the frit. When a cold solution is to be filtered, a different approach is taken. The entire frit assembly is cooled by "swabbing" the outside with a dry ice/ethanol slush(20). Inverting the frit assembly with the stopcock along the side arm closed and admitting a partial atmosphere of argon into the top half of the frit assembly will facilitate the process. A little forethought when assembling the frit, flasks and needle valve adapter (i.e. which side to put the needle valve adapter on) will save time later.

Washing a Solid on a Frit

To wash a solid which has been collected on a frit, the
flask containing the wash solvent (the filtrate in the
receiving flask or fresh solvent from a solvent pot
attached at a nearby port of the working section) is
cooled with a dry ice/ethanol slush bath and then
degassed. After closing the valve of the sidearm along
the side of the frit assembly, the slush bath is replaced
with a warm water bath and the top flask is cooled by
"swabbing" with slush. Condensation of the wash solvent
occurs on the upper half of the frit and upper flask the
percolator action of the solvent effectively washes the
solid. The resultant solution is forced back into the
bottom flask by replacing the warm water bath with a
slush bath. The valve of the side arm should be closed
during this process.

Procedures for Carrying Out Reactions in an NMR Tube

We routinely use high field NMR spectrometry to screen
reactions, to characterize compounds and to measure
reaction kinetics. By working in a sealed NMR tube it is
possible carry out reactions involving very air sensitive
materials an volatile reagents, as well as those which
require conditions of high temperature or moderate
pressure (0 - ca. 3 atm). The procedure for assembling
the reagents for a reaction in a sealed NMR tube is
described below for the reaction of $(C_5Me_5)_2ScCH_3$ with
2-butyne. In the inert atmosphere glove box, an NMR
tube(21) fused to a small diameter standard wall pyrex
tube and then to a 14/20 standard taper outer joint is
charged with a known amount of the solid starting
material. The NMR tube is then attached to a calibrated
gas volume such as the one shown in Figure 7. Deuterated
solvent (approximately 0.5 ml) is added either in the
glove box or by vacuum transferring from a NMR solvent
pot on the vacuum line. The NMR tube is then evacuated
while cooling at $-78^{\circ}C$, and the stopcock separating the
NMR tube from the calibrated volume above is closed.
2-butyne (which has been dried over 4 A molecular sieves)
is freeze-pump-thawed on the vacuum line, and a specified
pressure is admitted to the manifold and gas volume above
the NMR tube. Residual 2-butyne is returned to the
storage to its storage container by cooling. The known
amount of 2-butyne in the gas volume is then condensed
into the NMR tube $(-196^{\circ}C)$. The tube is now sealed with
a cool flame of a glass blowing torch while maintaining
cooling $(-196^{\circ}C)$ of the lower half of the tube. Sealed
NMR tubes should be allowed to warm behind proper
shielding in a fume hood, since condensed O_2 or excess
pressure can lead to an explosion. If the tube is to be
heated in the NMR probe, pre-testing of the tube strength
by heating briefly at at least $20^{\circ}C$ higher temperature
than the probe is highly recommended. If volatiles are

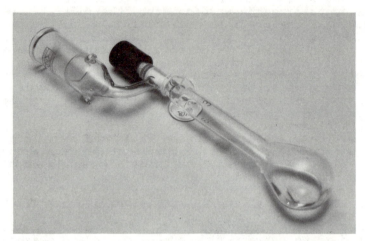

Figure 6.   Container for volatile reagents.

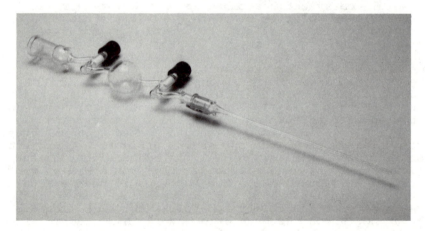

Figure 7.   NMR sample tube attached to calibrated gas
            volume assembly.

condensed into an NMR tube which exhibit greater than one atmosphere vapor pressure at the temperature of the probe, a preliminary calculation should be done to determine what the actual pressure inside the NMR tube will be during the experiment. For pressures approaching or greater than five atmospheres, use of medium walled or thick walled NMR tubes is recommended.

Thick Walled Reaction Vessels

Many of our reactions are carried out at high temperatures and/or moderate pressures (1-7 atm). A convenient way to do this type of reaction is to use a thick-walled reaction vessel, shown in Figure 8. Solid starting materials are added in the inert atmosphere glove box; solvents and gases are added on the vacuum line as described above. The glass vessel can then be heated in an oil bath for the appropriate length of time. Once the reaction is completed, excess pressure is released, for example by cooling the vessel and carefully venting it through a trap cooled at -196°C. The solution is transferred to a swivel frit assembly for work up, using either an inert atmosphere glove box or a syringe attached to polyethylene tubing, following the procedure described above for adding solutions to the swivel frit assembly. The second method is much quicker and is suited to solutions of thermally sensitive, moderately air sensitive compounds.

Use of a Toepler Pump

To quantify the amount of a noncondensable[22] gas such as H_2, CO, CH_4, or N_2 liberated during a reaction, a Toepler pump is employed. The basic operating principle of a Toepler Pump is the use of a mercury piston which is pushed up the upper cylinder by atmospheric pressure admitted to the lower chamber, and allowed to fall as the lower cylinder is evacuated. These procedures are alternated by the relay/control mechanism described earlier, and the gas collected on each cycle is trapped by a one way, float valve in the upper calibrated volumes. Our pumps use a a simple stainless steel ball bearing seated into a spherically ground portion of the valve. Toepler pumps of several different designs are commercially available; ours are of a hybrid design and are constructed in the Caltech glass shop. If the volume of the upper cylinder is 500 ml and the volume being pumped (the cryogenic traps, manometers, manifolds etc. back to the reaction flask is 1500 ml, 75% of the residual gas of the sample being collected will be left behind on each cycle. Thus, many cycles will be needed to reduce the amount of residual gas to a negligle value $((0.75)^n \longrightarrow 0$, where n is the number of Toepler pump cycles). An indication of the amount of residual gas is provided by the size of the bubble formed when the

mercury approaches the upper valve. The pressure in the
calibrated volume should be measured occasionally (by
allowing the Hg level to rise to the mark above the
valve). When no difference is detected from reading to
reading, gas collection may be stopped.

   To separate (hot CuO) combustible gases(23) from
noncombustibles, the sample is first collected and the
total amount of gas is measured. The stopcock leading
back to the cryoscopic traps is closed, the stopcocks
opening the pathway around the recirculating manifold
section are opened. The teflon needle valve leading back
to the last trap is opened only slightly so that
excessive pressures of gas are prevented from building up
at the inlet of the Toepler pump. The gas mixture is now
circulated by the Toepler pump through the heated CuO
section (320°C), the trap cooled to -196°C and back into
the calibrated section, etc. until all H_2 is oxidized to
H_2O and CO is oxidized to CO_2, both of which are trapped
out at -196°C. By closing the stopcock immediately above
the last section of the calibrated volume, the residual,
noncombustible gas (N_2 and/or CH_4) may be collected and
its pressure determined. By periodically repeating these
procedures it can be determined if more cycles through
the CuO tube are required. The residual gas is measured
and analyzed by infrared spectroscopy, gas chromatography
or mass spectrometry by withdrawing a sample into a gas
ir cell or gas sample bulb attached to the port at the
top of the Toepler pump section of the vacuum line. The
residual N_2/CH_4 is pumped out of the calibrated volumes.
The trap at -196°C is warmed to -78°C by replacing the
liquid nitrogen with a dry ice/ethanol in the dewar
surrounding it, and the CO_2 is now collected by the
Toepler pump. The amount of CO_2 is, of course, the same
as the amount of CO in the original sample. The
difference is the amount of H_2, now as H_2O in the -78°C
trap, in the original sample. Thus, by applying Dalton's
Law of Partial Pressures, one can easily quantitatively
determine the composition of H_2/CH_4, H_2/N_2, CO/CH_4,
CO/N_2, CO/H_2, $CO/H_2/CH_4$, and $CO/H_2/N_2$ binary and tertiary
mixtures, and if coupled with mass spectrometry to
quantify the proportions of methane in N_2, $CO/H_2/CH_4/N_2$
quaternary gas mixtures.

Molecular Weight Determination for Non-Volatile
Organometallic Compounds

Of all the methods that are available for determining the
molecular weight of an organometallic compound, the most
straightforward is undoubtedly that developed by
Signer(24). This method is based on the simple idea that
the vapor pressure of an ideal solution is proportional
to the concentration of the solute (Raoult's Law). The
major advantages of this technique is that it does not
require any sophisticated equipment, only about 0.05 g of
the unknown compound are needed and it can be suited to

extremely air sensitive compounds with the adaptations described below. The limitations of this method are that the sample must be stable in solution at room temperature for 3-7 days and the accuracy of this method is ±10% at best. The molecular weight apparatus consists of two graduated capillary tubes (approximately 10 ml in volume, constructed from sections of a Mohr pipet) each sealed to a round bulb. The two bulbs are connected to a short glass tube with solvent seal connectors. In the middle of the glass tube, there is a stopcock through which the entire assembly can be evacuated. A diagram of the apparatus is shown in Figure 9. The procedure for determining the molecular weight of an unknown air sensitive compound is outlined below.

In the glove box, 0.010-0.015 g of the standard (azobenzene or ferrocene are good choices) is carefully weighed out (± .1 mg) and placed in one bulb. Similarly, an amount of the unknown is accurately weighed out and placed in the other bulb. A solvent which is very inert toward the sample (benzene, cyclohexane, tetrahydrofuran) is added (ca. 0.5-0.8 ml) to each bulb and the apparatus is assembled. The bulbs are cooled to -78°C, and the apparatus is evacuated on the vacuum line. The assembly is then allowed to come to room temperature and placed (with the solvent in the bulbs) in a location free from drafts, away from direct sunlight or room light (a cardboard box or large pot work nicely). The volume change should be periodically monitored by tilting the assembly so that the solutions flow into the volumetric portions of the apparatus. The volume of each solution will change by transfer of solvent vapor until the vapor pressures of the two solution equilbrate, usually requiring 3-7 days. The experimental molecular weight is determined by the following relationship:

$$MW_x = \frac{(mg_x)(MW_s)(ml_s)}{(mg_s)(ml_x)}$$

where mg_x = weight of unknown in mg
mg_s = weight of standard in mg
MW_x = molecular weight of unknown
MW_s = molecular weight of standard
ml_x = volume of standard solution
ml_s = volume of unknown solution

Acknowledgments   The authors wish to thank members of the Bercaw group, past and present, who have contributed to the development of the procedures and equipment described in this chapter. JEB wishes to thank Professor Hans H. Brintzinger (presently at the University of Konstanz, West Germany) for inspiring many of the original concepts when the two of us were at the University of Michigan.

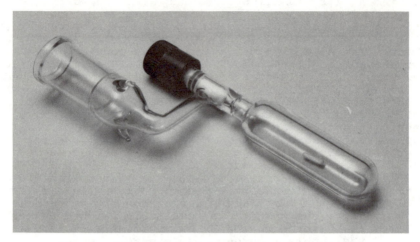

Figure 8.   Thick walled reaction vessel.

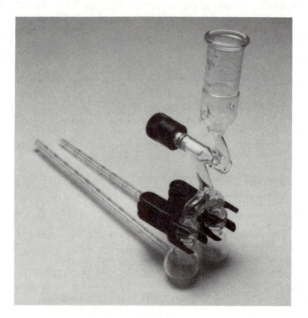

Figure 9.   Apparatus for molecular weight
            determination.

REFERENCES AND FOOTNOTES

1. Because it operates with a Hg column, the McLeod gauge does not read the partial pressure of mercury (ca. 10^{-3} torr) which continually fills those portions exposed to liquid Hg.

2. The accurate measurement of pressure requires use of a manometer. The bubbler is less accurate since the level of the pool of mercury at the base varies slightly as the column height varies.

3. Our group has invested heavily in Fischer Porter brand 4 mm needle valves. These provide fine control when they are opened, and are thus preferable for those sections of the system where control is needed, for example, for solvent removal and at the points of entry to the cryogenic trap and Toepler section. In recent years less expensive, but equally effective teflon needle valves have become available.

4. The base of the Toepler pump and the bases of the mercury bubblers should be placed in plastic cups and secured with plaster of paris to a depth of 3 or 4 inches. Thus, mercury spillage is avoided is the glass base is broken.

5. The pathway of this recirculating section should have as little volume as possible out of the circuit, since gases must flow freely to prevent them from "hiding in corners" from the reactive CuO.

6. A simplified ideal gas formula is used to calculate the mmols of gas in the volumes: n (mmols) = p (torr)$\cdot$V (ml)/1.85×10^4 at 23.5°C.

7. The circuit diagram is available on request.

8. Brown, T. L.; Dickerhoof, D. W.; Bafus, D. A.; Morgan, G. L. Rev. Sci. Instruments 1962, 33, 491-492.

9. Care should be taken not to allow the aluminum foil to contact the electrical leads!

10. Since Ar has only ca. 100 torr of vapor pressure at liquid nitrogen temperature, a dangerous buildup of liquid Ar will occur if the pressure of Ar exceeds this value while a trap, flask, NMR tube, etc. is being cooled with liquid nitrogen. Similarly care must be taken to not allow the pressure to exceed one atmosphere when cooling a part of the line with N_2 as the inert gas.

11. The purity of commercially available carbon monoxide (e.g. Matheson) is normally sufficient for our experiments.

12. Marvich, R. H.; Brintzinger, H. H. J. Amer. Chem. Soc., 1971, 93, 2046-2048.

13. For solvent pots of greater than 500 ml in volume, it is recommended that they be taped or otherwise treated to prevent flying glass should the flask implode. This advisory note applies whenever large glassware is evacuated.

14. Heating the solvent pot above room temperature will
    only slightly increase the rate of transfer, and has
    the undesirable effect of causing solvent
    condensation along the path of the transfer.

15. Thicker Apiezon H grease is used to compensate for
    the poorer fit of standard taper joints and to
    reduce the amount of solvent wash out. The joints
    are greased only along the top half of the ground
    glass section to minimize contamination from grease;
    however, we have never noted deleterious effects of
    Apiezon greases on our compounds.

16. If the high vacuum working station is used the
    bubbler/manometer may be accurately used as a
    manometer since the pools of Hg are interconnected
    (see Figure 1).

17. If a very reactive, noxious or poisonous reagent is
    used, the residual reagent should be trapped out
    into one of the removable traps, and the reagent
    should be carefully transferred to the fume hood
    where it is allowed to warm slowly.

18. A section of a Mohr pipet or graduated centrifuge
    tube sealed to a 14/20 standard taper outer joint
    have proven to be quite satisfactory.

19. Caution: the reaction flask may be cooled only to
    -78°C; cooling to liquid nitrogen temperature will
    lead to dangerous buildup of liquid Ar. If the
    reaction flask must be cooled to liquid nitrogen
    temperature, N_2 should be used as the inert gas and
    great care must be exercised to avoid condensation
    of liquid nitrogen.

20. Care should be taken to avoid cooling the teflon
    valves. The much different coefficients of thermal
    expansion for teflon and glass results in leakage
    when the valve is cooled much below room
    temperature.

21. We have found that Wilmad (507 pp) NMR tubes provide
    adequate resolution for routine high field NMR
    spectra.

22. A noncondensable gas is operationally defined as one
    that has more than ca. 0.01 torr pressure at liquid
    nitrogen temperature.

23. Care should be taken to avoid O_2, even in small
    amounts, as a component of these gas mixtures. An
    explosive mixture with H_2, CO or CH_4 could easily be
    produced and detonated by sparks at the electrodes
    of the relay or on contact with the hot CuO.

24. E.P. Clark, Ind. Eng. Chem., Anal. Ed. 1941, 13,
    820..

RECEIVED July 31, 1987

Chapter 4: Application 1

Assay of Ligand-Derived Gases
Following Outer-Sphere Oxidation

Eric G. Lundquist and Kenneth G. Caulton

Department of Chemistry, Indiana University, Bloomington, IN 47405

A method for the quantitative evolution and estimation of a ligand-derived gas is described. Dissociation of the gas (CO, H_2, CO_2, N_2, etc.) is promoted by outer-sphere oxidation of the compound of interest using $[Fe(bipy)_3](PF_6)_3$ in acetonitrile solvent. Examples described include the analysis of CO content in $Fe_3(CO)_{12}$ and $Co_2(CO)_8$.

An extremely valuable development in the field of metal carbonyl chemistry was the publication of an analytical method for the determination of the number of carbonyl ligands contained in a complex ([1],[2]). This became increasingly important with the discovery of metal carbonyl clusters, since the weight percent carbon between various molecular formulae varies little when the atom ration C/M is no longer integral. Thus, the rhodium carbonyls $Rh_2(CO)_8$, $Rh_4(CO)_{12}$, $Rh_6(CO)_{16}$ have C/Rh ratios (4, 3, and 2.67) which tend towards a limit and correspondingly similar percent carbon values. Classical elemental analysis is of essentially no use in establishing the number of carbonyl ligands in the large anionic clusters of the Longoni/Chini group ([3]). For example, the recent examination of an apparent structural isomer of $Ru_3(CO)_{12}$, while it gave acceptable percent carbon (calc., 22.54, found, 22.04), would perhaps benefit from direct and accurate measurement of a CO content ([4]). Similarly, identification of the prodigious array of osmium carbonyls of high nuclearity ($Os_2(CO)_9$, $Os_3(CO)_{12}$, $Os_5(CO)_{16}$, $Os_5(CO)_{19}$, $Os_6(CO)_{18}$, $Os_6(CO)_{20}$, $Os_7(CO)_{21}$) relies little, if at all, on elemental analysis ([5]. In each instance, the modern phase of metal carbonyl chemistry has relied on x-ray diffraction to determine not only bond lengths and angles, but even molecular formula; modern x-ray diffraction has become a technique of qualitative as well as quantitative analysis. Yet, there are cases where carbonyl ligands comprise such a small fraction of the total x-ray scattering power that even the number of CO ligands cannot be reliably detected ([6],[7]). In addition, there are compounds

0097–6156/87/0357–0099$06.00/0

in which disorder frustrates accurate crystallography, as well as compounds which ''refuse'' to crystallize.

Against this background, we have sought an analytical procedure which reliably (i.e. quantitatively) converts ligands to gaseous molecules which are then quantitated by standard PVT measurements using a utility vacuum line (8). Since all ideal gases occupy the same molar volume regardless of molecular weight, this procedure is particularly advantageous for lightweight ligands, where gravimetric procedures suffer their most acute inaccuracies. Thus, while the method has been introduced in the above paragraphs for carbonyl ligands, its advantages are greatest for the lightest of all ligands, hydride (leading to H_2), which also suffer from poor visibility to x-rays because of their low scattering power (9). Extension of the methodology to other ligands which might be released as a gas (e.g. N_2, CO_2, C_2H_4, $C_2O_4^{2-}$, BH_4^{0}, CO_3^{2-}, etc.) is left to the imagination of the reader.

Method

The method of evolution of a ligand as a gas relies on the principle that removal of electrons from a complex will weaken metal-ligand bonds. While this is obvious in the case where the oxidation removes metal-ligand bonding electrons (e.g. the d^0 complexes WMe_6 or $ReH_7(PR_3)_2$), it can also be true when the electrons are more localized on the metal. Thus, to the extent that the metal-carbonyl bond depends on back bonding, removal of the d-electrons from $M_m(CO)_n$ will weaken the M/C bond. Said in another way, carbonyl ligands become improbable for metals in high formal oxidation states. For hydride ligands, oxidation of a metal/hydride complex will achieve a state where the reducing power of the hydride ligand is sufficient to cause internal electron transfer to the (oxidized) metal, perhaps generating a coordinated H_2 molecule (Equation 1). Coordinated molecular H_2 is known to be kinetically labile, just as

$$L_mM^n(H)_n \xrightarrow{-Xe^-} L_mM^{n+X}(H)_n^{X+} \longrightarrow L_mM^{n+X-2}(H)_{n-2}(H_2)^{X+} \qquad (1)$$

are the CO ligands in an oxidized (especially a paramagnetic) carbonyl complex. Thus, evolution of gaseous ligand may be expected to ensue in the presence of an abundance of incoming ligand chosen to stabilize a higher metal oxidation state. We have found acetonitrile to be a satisfactory solvent/ligand, but this choice might be productively altered by others as their particular system dictates. Application of the method to polyhydride compounds has been described in some detail, as has the detrimental influence of protic nucleophiles as solvents or impurities (9).

The choice of oxidant remains. Our reasoning has been that we wish to avoid inner sphere electron transfer reagents since the group (halide, oxalate, etc.) which bridges during electron transfer can also stabilize the oxidized unknown against quantitative gas evolution. Examples include conversion of $Fe(CO)_5$ to $Fe(CO)_4I_2$.

Our work to date has employed $Fe(bipy)_3^{3+}$ (bipy = 2,2'-bipyridine), with $E^o(MeCN) = 1.21$ v. as well as the weaker oxidant Cp_2Fe^+ (E^o = +0.55 v) both vs. S.C.E. Each has the additional advantage that it undergoes a vivid color change on reduction (from deep blue, in each case, to deep red or pale yellow, respectively). Here again, the reader is encouraged to seek alternative (superior) chemical oxidants according to the dictates of the system under study. A more general procedure would be to use an electrode to effect "oxidative stripping" of ligand as a gas, since the question of oxidation potential is more readily under the control of the experimenter.

Procedure

The experimental procedure consists of loading a vacuum-tight reaction vessel (typical dimensions, 3 X 20 cm) with a magnetic stir bar, 5-6 moles of oxidant (e.g. $[Fe(bipy)_3](PF_6)_3$) per mole of unknown and a small (1 mL) open-ended vial containing a weighed amount of the compound to be analyzed. For a detailed synthesis of this oxidant, see ref. 9. The amount of test compound is chosen so that the number of mmoles of gas expected is in the range where accurate quantitation is possible with the calibrated utility vacuum line available. We typically require 0.5-2.0 mmole of gas, but capillary tubing in the vacuum line would permit consumption of less compound in cases where one has available only tenths of mmoles of compound.

Dried, freeze-thaw degassed solvent (e.g. acetonitrile, 5 mL) is then vacuum transferred into the evacuated, cooled (-196°C) reaction vessel, which is then closed and allowed to warm to 25°. Our experience has been that gas evolution is visually evident at this temperature. While gas evolution is often complete within one hour, we have experienced cases where up to 24h is required. The most objective procedure would be to let the oxidation proceed with the reaction vessel open to a manometer, and use the cessation of pressure changes to signal completion of the reaction.

Separation of a noncondensible gas like CO or H_2 from the solvent is effected simply by carefully opening the reactor (at 25°) to a cold (-196°) spiral trap (of approximate volume 100 mL). After completion of vacuum transfer of acetonitrile into the sprial trap, the volatile gas is Toepler pumped (10) into the calibrated region of the vacuum line until further strokes of the pump give no change in pressure. The number of mmoles of gas is determined from the calibration of the vacuum line.

Two final checks should be performed. Following quantitation, the identity of the evolved gas should be established using appropriate methods (mass spectrometry, infrared or Raman spectroscopy, etc.). The solid residue in the reactor from the oxidation should also be probed with vibrational and/or NMR spectroscopy to establish that none of the ligand of interest remains. In the event that some ligand remains following oxidation, quantitation of the ligand of interest is still possible if the identity and yield of the oxidized compound can be established.

Examples. In a typical example, 20.0 mg of an iron carbonyl with 33.27 weight percent iron (i.e. 0.119 mmol Fe) was combined with 230 mg (0.714 mmol) $[Fe(bipy)_3](PF_6)_3$ and 5 mL CH_3CN was added by vacuum transfer. Gas evolution began near 25°. After 1.5h, the solvent was removed from the solution by vacuum transfer and the noncondensible gas (CO) was Toepler pumped into a calibrated volume for pressure measurement. The average of four pressure measurements gave 0.47 mmol CO, for a CO/Fe ratio of 3.94. The iron carbonyl is thus $[Fe(CO)_4]_n$ where, in this case, n = 3.

In a second example, 25.0 mg of a cobalt carbonyl containing 34.48 weight percent cobalt (0.146 mmol Co) was combined with 420 mg $[Fe(bipy)_3](PF_6)_3$. Acetonitrile (5 mL) was added by vacuum transfer. Separation and quantitation as above gave 0.60 mmol of CO evolved, for a CO/Co ratio of 4.10. The empirical formula of the compound is thus determined to be $Co(CO)_4$; the compound employed was $Co_2(CO)_8$.

Extensions of the Method

The oxidatively-induced ligand elimination described here is also useful in quantitating the extent of incorporation of stable isotope in some exchange reaction. Quantitative evolution of $^{12,13}CO$ or $(H,D)_2$ would then be followed by mass spectral or vibrational spectroscopic determination of isotopic composition. The whole subject of oxidatively-induced reductive elimination (11,12) would suggest that methyl, or even higher alkyl ligands might be similarly quantitated.

Literature Cited

1.   Sternberg, H. W.; Wender, J.; Orchin, M. Anal. Chem. **1952**, _24_, 174.
2.   Hieber, W.; Sedlmeier, J. Chem. Ber. **1954**, _87_, 25.
3.   Chini, P. Adv. Organometal. Chem., _14_, 285.
4.   Hastings, W. R.; Baird, M. C. Inorg. Chem. **1986**, _25_, 2913.
5.   Lewis, J.; Johnson, B. F. G. Pure Appl. Chem. **1982**, _54_, 97.
6.   Ceriotti, A.; Demartin, F.; Longoni, G.; Manassero, M.; Marchionna, M.; Piva, G.; Sansoni, M. Angew. Chem. Int'l. Ed. **1985**, _24_, 697.
7.   Chini, P. J. Organometal. Chem. **1980**, _200_, 37.
8.   Shriver, D. F. In The Manipulation of Air-Sensitive Compounds; McGraw-Hill Book Co., New York, 1969.
9.   Lemmen, T. H.; Lundquist, E. G.; Rhodes, L. F.; Sutherland, B. R.; Westerberg, D. E.; Caulton, K. G. Inorg. Chem. **1986**, _25_, 3915.
10. See contribution of J. Bercaw in this volume.
11. Saez, I. M.; Meanwell, N. J.; Nutton, A.; Isobe, K.; Vazquez/deMiguel, A.; Bruce, D. W.; Okeya, S.; Andrews, D. G.; Ashton, P. R.; Johnstone, I. R.; Maitlis, P. M. J. Chem. Soc. Dalton Trans. **1986**, 1565.
12. Kochi, J. K. In Organometallic Mechanisms and Catalysis; Academic Press, New York, 1978; pp. 282, 351-352.

RECEIVED September 1, 1987

Chapter 4: Application 2

Semiautomatic Gas Titration Device

R. E. King III

Specialty Chemicals Division, Union Carbide Corporation, Tarrytown Technical Center, Tarrytown, NY 10591

The semi-automatic gas titration device is a useful tool for measuring the rate, as well as the extent, of a gas consuming reaction. In terms of measuring the rate, or completeness, of a gas consuming reaction, this device is a versatile, precise, semi-automatic, time saving tool far exceeding the capabilities of a mercury manometer equipped with a meter stick.

In principle, the semi-automatic gas titration device works by monitoring the partial vacuum created by a chemical reaction that consumes a gaseous substrate within a closed system.[1,2] As the reaction progresses, the consumed gas is continually replenished in small uniform aliquots through a parallel set of solenoids connected to the reaction vessel and to a source of fresh reactant gas.

A general schematic of the titration device is shown in Figure 1. The partial vacuum created by the gas consuming reaction is used to move mercury between two platinum electrodes in an electronically modified mercury manometer. When the mercury level has fallen from the top electrode, E1, to just below the second electrode, E2, a process cycle controller activates the solenoids, B1 and B2. This solenoid action temporarily closes off the manometer to the reaction vessel and opens it to a source of reactant gas. Consequently, the loss in pressure is compensated by the introduction of fresh reactant gas to the manometer. The pressurization of the manometer forces the mercury level from below E2 back up to E1. When the mercury level reaches E1, the process cycle controller deactivates the solenoids. The

0097–6156/87/0357–0103$06.00/0

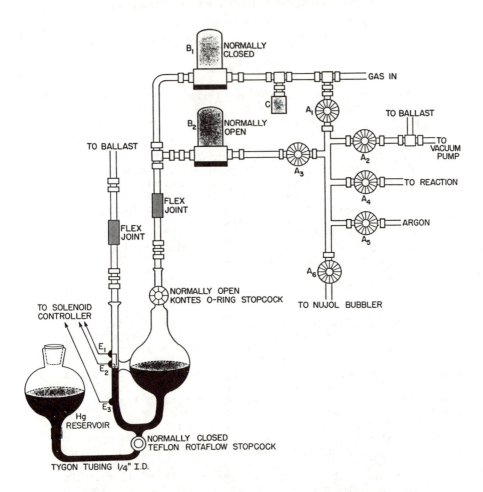

Figure 1.   Schematic of the semi-automatic gas titration device, depicting the arrangement of the electronic manometer, solenoids, and manifold.

manometer is subsequently re-opened to the vessel where
the reaction is taking place. This cycle repeats itself
until the limiting reagent is consumed.

Description of the Apparatus

The titration device and its components are constructed
of gas-tight stainless steel tubing, fittings and
valves. The manometer and the reaction flask; however,
are constructed of thick walled glass equipped with O-
ring fittings and greaseless stopcocks. As can be seen
in Figure 1, the central manifold is constructed in such
a way that the valves to the reaction gas, high vacuum
system, mercury manometer, reaction vessel, inert gas,
and pressure release bubbler, A1 - A6 respectively, are
configured in a compact arrangement and therefore, are
readily accessible for quick manipulation.
    The titration device and its various components,
are attached to and supported by a sturdy rod type rack,
preferably on top of a waist high table. The use of
rigid attachments of the components to the rack is also
preferred over conventional three fingered clamps, in
that these clamps require periodic tightening,
potentially leading to an desireable buildup or
redistribution of stress in the system. The final
arrangement of the apparatus and its various components
will be up to the individual experimentalist, based on
space limitations; however, it is most convenient to
have the system designed so that individual components
of the apparatus are somewhat separated and can be
serviced without jeopardizing the mechanical integrity
of the other components.
    The most important component of the semi-automatic
gas titration device is the electronically fitted
mercury manometer. A detailed schematic is shown in
Figure 2. It is important to point out the use of glass
to metal seals which allows the glass manometer to be
firmly attached, via flexible stainless steel tubing, to
the rest of the titration device for an air-tight
connection. The mercury manometer is also supported by
a sturdy, Tygon padded, ring clamp (not shown). The
steel flex joints are used to connect the heavy, and
potentially breakable, glass mercury manometer to the
rest of the system. The flex joints and padded ring
clamp help to relieve some of the strain and rigidity
inherent in the overall system. The platinum
electrodes, E1, E2 and E3, are connected through the
wall of the manometer tubing, and are also protected by
glass to metal seals. These electrodes are connected to
the process cycle controller which controls the
activation and deactivation of the solenoids. The use
of platinum electrodes is favored due to their
resistance to oxidation. The manometer is also fitted
with a mercury reservoir so that the gas in the

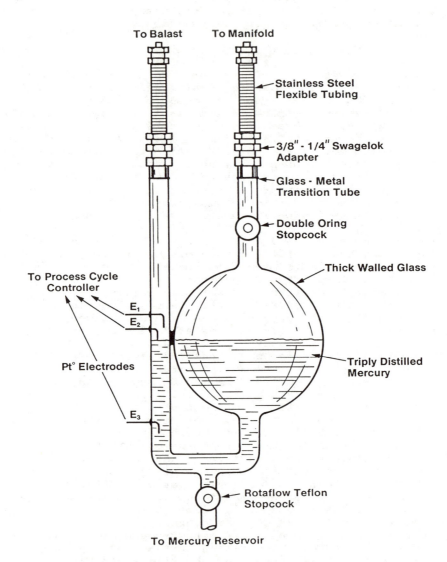

Figure 2.   Schematic of the electronic mercury manometer.

manometer can be efficiently purged, by completely
filling the manometer with mercury from the reservoir,
if the gas in the manometer is inadvertantly
contaminated by air or inert gas.

   The right side of the mercury manometer, as shown
in Figure 1, is connected to the central manifold
through the normally open solenoid, B2, and to the
reaction gas through the normally closed solenoid, B1,
through a T-connection. The left side of the manometer,
also shown in Figure 1, is connected to a ballast tank.
The ballast tank, shown in greater detail in Figure
3(a), is used to adjust and maintain the pressure at
which the gas consuming reaction is to be performed,
typically atmospheric pressure. However, the ballast
tank can be particularly useful for varying the pressure
at which the gas consuming reaction is to be performed.
Consequently, rate studies can be performed in order to
determine the reaction order with respect to the
reactant gas.

   The ballast tank should be large enough, 10-15
liters, as not to require additional work to compress or
decompress the gas inside the manometer while the
mercury moves up and down between the two platinum
electrodes. The ballast tank is fitted with connections
to a source of inert gas for pressurizing the tank, the
vacuum system for depressurizing the tank, the mercury
manometer, and an accurate pressure or gauge for
measuring pressure in the ballast tank. Using inert gas
or vacuum, the pressure inside the tank can be adjusted
to the desired setting, thereby dictating the pressure
at which the reaction will be performed. Typical
pressures that can be used with this device range from
0.25 to 2.5 atmospheres, gauge pressure. Appropriate
shielding and precautions must be taken when operating
above or below atmospheric pressure. The ballast should
also be well insulated to minimize pressure fluctuations
due to changes in room temperature.

   IMPORTANT: It should be obvious that the other
side of the mercury manometer must be similarly
pressurized or evacuated. High or low pressure
experiments should only be performed by those well
acquainted with the system and have run numerous
experiments at atmospheric pressure.

   The reaction vessel, shown in Figure 3(b), is
constructed of thick walled glass. The reaction flask,
i.e., inner shell, is a Mortonized flask fitted with an
efficient reflux condenser, a sidearm/stopcock for
introducing reagents or withdrawing samples, an outer
shell for continuously passing a temperature regulated
liquid around the inner flask in order to maintain
constant temperature, and a fan blade stir bar to insure
rapid introduction of the gas into the reaction medium.

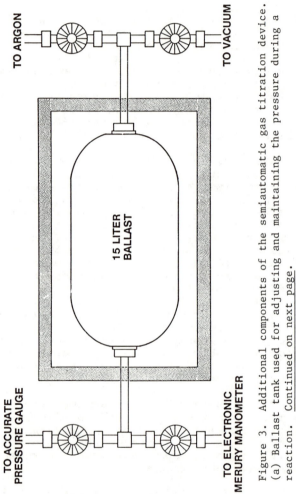

Figure 3. Additional components of the semiautomatic gas titration device. (a) Ballast tank used for adjusting and maintaining the pressure during a reaction. Continued on next page.

b

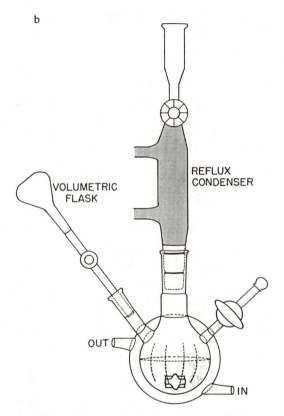

Figure 3. <u>Continued</u>. (b) Reaction vessel fitted with reflux condenser and volumetric flask. <u>Continued on next page</u>.

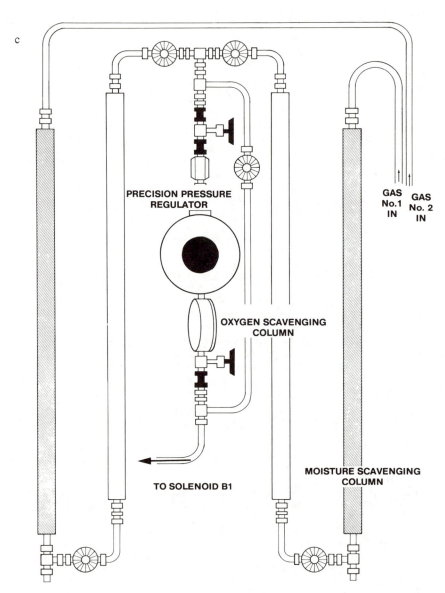

Figure 3. Continued. (c) Purification columns for the reactant
gases.

The flask is also equipped with a sidearm O-ring joint, to which a volumetric flask, fitted with a greaseless stopcock can be attached. The use of the volumetric flask is an efficient and precise way of introducing the reagents and solvent into the reaction vessel. This is also particularly useful when working with air sensitive materials, in that the reaction flask and volumetric flask can be charged with the reagents in an inert atmosphere glove box.

Another component of the gas titration device is the purification system for the reactant gas, shown in Figure 3(c). The reactant gas is first passed through a column that scavenges moisture, and then through a column that scavenges oxygen. The gas pressure is regulated by a standard pressure gauge off the tank, but should also be passed through a precision pressure regulator for use in obtaining reliable data during low pressure studies. As can also be seen in Figure 3c, a bypass for the precision pressure regulator is incorporated for high pressure studies, where the tank regulator should suffice. A second reactant gas train is configured in a similar fashion, and can be used as a backup during purification of the first set of columns, or it can be used for a different reactant gas; e.g., deuterium instead of hydrogen.

The schematic of the process cycle controller is shown in Figure 4. The solenoids are in the deactivated mode while the manometer is open to the reaction vessel. When the mercury level drops below electrode E2, the solenoids are activated and the manometer is shut off to the reaction vessel and opened to a source of fresh reactant gas. The rate of the refill, about 5 seconds, can be controlled by the metering valve, C, shown in Figure 1. When the mercury level reaches E1, the solenoids are deactivated, the counter is reset, and the manometer is opened to the reaction vessel once again.[3]

The data, which consists of time intervals between refill cycles, can be recorded on a simple strip chart recorder by sending an electrical pulse to the recorder. Each time interval is correlatable to the consumption of a uniform aliquot of reactant gas. The aliquots of gas consumed are directly related to the reagent being chemically transformed, e.g., an olefin being reduced by hydrogen. Therefore, the time between refill cycles and the amount of gas consumed can be graphed to give a plot of moles of reagent consumed versus time. Based on the characteristics of the plot, the experimentalist can determine the reaction order with respect to the reagent being consumed. The means by which the data is collected is again up to the experimentalist and the resources available. The author's preferred method was to have the time intervals printed out on a strip of paper tape on a printer attached to the process cycle controller. This data then served as input to a

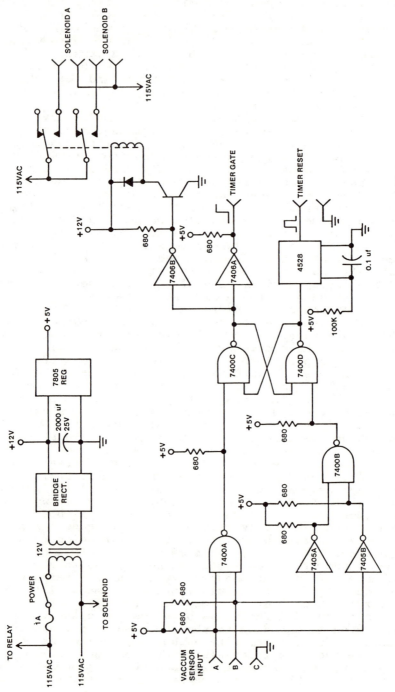

Figure 4. Electrical diagram of the process cycle controller.

computer program which solved for the reaction order,
based on the reagent being transformed. In a more
sophisticated approach, the data can be directly input
to a dedicated lab computer and the experiment can be
analyzed as the reaction proceeds.

Procedure

As a representative example for the implementation of
this device, consider the hydrogenation of a
non-volatile olefin using an air sensitive
organometallic catalyst. The reaction flask components
are oven dried and then transferred into an inert
atmosphere chamber or dry box. The catalyst is
accurately weighed out in the dry box and dissolved in a
prescribed amount of solvent. The catalyst solution is
transferred via syringe into the volumetric flask to the
mark, and the stopcock closed. The olefin is weighed
and charged into the reaction flask. The reaction flask
is fitted with the volumetric flask, the fan blade
magnetic stirring bar, and the reflux condenser. The
components of the reaction vessel are secured with
rubber bands and then removed from the dry box with all
stopcocks closed. The reaction vessel is then be
attached to the titration device using quick connect
fittings capable of withstanding both high and low
pressures.
     The reaction vessel is evacuated, through valve A2,
and then refilled with inert gas, through valve A5,
three times. Once the reaction vessel in under an inert
atmosphere, the excess pressure of inert gas can be
released, through valve A6, a Nujol bubbler, or
preferably, a Firestone valve. The stopcock on the
volumetric flask is opened and the catalyst solution is
allowed to drain into the flask. Upon complete mixing
of the reagents, the stopcocks on the reaction vessel
are closed off and the solution is frozen with a stream
of liquid nitrogen which is passed between the inner
reaction flask and the outer shell. The gas remaining
above the frozen solution is then removed by opening the
flask to the high vacuum system, valve A2. When the
system reaches high vacuum, the reaction vessel is
closed off to the vacuum system and the solution is
allowed to thaw.
     After three freeze/pump/thaw cycles, the reaction
vessel is fitted with hoses supplying liquid from a
constant temperature bath and allowed to equilibrate to
the desired temperature. Concurrently, coolant is
supplied to the reflux condenser. Upon temperature
equilibration, the reaction vessel valve, A4, is opened
and the vessel is refilled with hydrogen, through valve
A1. Again, excess pressure is released through the
pressure release bubbler. The reactant gas, valve, A1,
and the pressure release bubbler valve, A6, are then

closed and the reaction vessel is opened to the
manometer by carefully opening valve A3. At this point,
the pressure between the manometer and the reaction
flask is very carefully adjusted with reactant gas or
partial vacuum so that the mercury level in the
manometer drops below electrode E2, which causes the
timer of the process cycle controller to be reset. When
the counter starts, the stirring is started and the
reaction begins. The reaction is over when the mercury
in the manometer stops moving or the time shown on the
timer is approximately three times that of the
previously recorded data point.

If the olefin is volatile, the procedure is
followed as described above except that the olefin is
not initially charged into the reaction vessel. After
the third freeze/pump/thaw cycle the reaction vessel is
refilled with hydrogen gas, through valve A1. Excess
pressure is released through the pressure release
bubbler, A6. The hoses from the constant temperature
bath are connected to the flask, and the condenser hoses
to the reflux condenser. After reaching thermal
equilibrium, the reaction is initiated by the injection
of the olefin, via an ultra precision syringe, through
the sidearm/stopcock fitted with a septum. The stopcock
is closed after the injection to assure against any loss
of gas. The procedure continues by shutting valves A1
and A6, opening valve A3, resetting the counter by
appropriately adjusting the pressure in the manometer,
and finally turning on the stirrer.

As a control experiment it is useful to calibrate
the titration device with a well documented
hydrogenation reaction. In this way the amount of
hydrogen per refill aliquot can be determined.[4] This
gives the experimentalist an idea of the number of
aliquots of reactant gas a particular reaction will
theoretically consume.

Acknowledgments

The author wishes to recognize the previous efforts of
Professor M.F. Hawthorne, J.J. Wilczynski, M.S. Delaney
and B.S. Anfield in designing and building the original
prototypes of this device, and J. Henigman of the
University of California, Irvine, Electronics shop for
designing and supplying the plans to the process cycle
controller.

References

1) This instrument is a modified version of a device
described by: (a) Sweet, Edward M., Ph.D. Thesis,
Michigan State University, 1976. (b) Larson, D.G.
Anal. Chem., 1971, 45, 217.

2) (a) Behnken, P.E.; Belmont, J.A.; Busby, D.C.;
Delaney, M.S.; King III, R.E.; Kreimendahl, C.W.;
Marder, T.B.; Wilczynski, J.J.; and Hawthorne, M.F.; J.
Am. Chem. Soc., 1984, 106, 3011-3025;
(b) ibid., 7444-7450.

3) The electrical diagram of the process cycle
controller, see Figure 4, shows the logic diagram used
to implement the control of the solenoids as well as an
external timer. When the vacuum increases and the
mercury level drops below the platinum electrode E2
(Sensor B) the output on 7400A will go low setting the
latch (7400 C & D) thereby activating the solenoids. As
the reactant gas refills the manometer, the mercury
level will make contact with the platinum electrode E2
(Sensor B), however, the latch will not change state
because electrode E1 (Sensor A) has not made contact
remaining a logic high. When both Sensors A and B make
contact, the 7400B output will go to logic low resetting
the latch. The 7405 provides a counter gate for a timer
which displays the time between Sensors A and B making
contact. The 4528 is a monostable multivibrator which
provides a short pulse to reset the timer.

4) Jardine, F.H. Prog. Inorg. Chem., 1981, 28, 63.

RECEIVED July 31, 1987

Chapter 5

Multipurpose Vacuum Line Designs for Use in Synthetic Organometallic Chemistry

Andrea L. Wayda, Patricia A. Bianconi, and James L. Dye[1]

AT&T Bell Laboratories, Murray Hill, NJ 07974

General design requirements for high vacuum line systems appropriate for use in synthetic organometallic chemistry are presented. Specific design considerations impacting on pumping station construction, vacuum gauge selection and specialty vacuum line features are then developed and discussed in relation to the modular, greaseless single and double manifold vacuum line systems which we have successfully constructed and used in our own laboratory.

The concept of 'vacuum' is difficult to define for a general audience. Depending on the scientist and the experiment, the same term is interchangeably used to denote pressures as high as several hundred torr and as low as 10^{-10} torr. Therefore, in writing a review article on vacuum line designs and techniques, it is prudent to acknowledge the co-existing meanings of the central concept and to define it carefully at the outset. This article confines itself to a discussion of the vacuum regime normally understood by synthetic organometallic chemists to encompass the high vacuum range extending from ca. 10^{-4} - 10^{-6} torr. However, in the discussion of roughing systems and Schlenk line designs, medium vacuum ranges (10^{-2} - 10^{-3} torr) will also be considered.

OVERVIEW

This review describes modular greaseless vacuum line systems which we have constructed in our laboratory for the synthesis and manipulation of air and moisture sensitive organometallic compounds. The discussion of these systems reviews important features of vacuum

[1]Current address: Department of Chemistry, Michigan State University, East Lansing, MI 48824

0097–6156/87/0357–0116$06.00/0
© 1987 American Chemical Society

line systems (as they apply to our synthetic needs) and is divided into four sections: Pumping Stations, Vacuum Gauges, Vacuum Lines, and Specialty Designs. Wherever possible, discussions of equipment will include recommended hardware with which the authors have had direct experience. Normally, several options will be presented, particularly where budgetary constraints may eliminate certain selections. Reference to specific equipment, price information (current as of June 1987), usage hints, etc. will be found in the footnotes. It should be stressed, however, that vacuum line designs are highly personal and idiosyncratic by nature. Other chapters in this volume will discuss approaches to the same end. In the sections below we describe and explain the design choices we have made so that the reader may choose the vacuum/Schlenk line system which best serves his or her experimental needs, whether described here or elsewhere in this book.

GENERAL DESIGN PHILOSOPHY

In constructing our systems, we have been influenced by several overarching design considerations. We have built systems which are:

1.) **COMPACT.** This feature allows the lines to be assembled and manipulations to be conducted entirely within a fume-hood thus protecting the laboratory environment and the researcher from noxious chemicals and implosion/explosion hazards.

2.) **MODULAR.** In contrast to the elegant permanent vacuum line installations described by Burger and Bercaw, we prefer our manifolds to function as high vacuum templates upon which specialty apparatus is constructed on an 'as needed' basis. This is accomplished by connecting specially designed apparatus (vide infra) into a functional unit using Cajon Ultra-Torr Unions and flexible stainless steel tubing. This arrangement is versatile and best suits our changing laboratory needs and the space constraints discussed above.

3.) **LOW MAINTENANCE.** The systems are designed to be grease-free, thereby eliminating the tedious regreasing and regrinding of stopcocks associated with traditional vacuum line systems. Glass valves are used in place of stopcocks, and valve bodies are constructed of Teflon fitted with O-rings compatible with the solvents used in that line. Line connections are made with O-ring joints, Ultra-Torr Unions or standard ISO flanges. This type of construction (combined with the modular nature of the components) allows the lines to be easily disassembled for cleaning, in their entirety, as needed.

THE PUMPING STATION

Any discussion of high vacuum line systems must begin with the

source of vacuum, the combination of forepump and diffusion pump
generically referred to as the pumping station. This is the heart
of any high vacuum line system. Since it is obvious that the best
obtainable vacuum in any system occurs at the throat or inlet of the
diffusion pump, the station must be designed to compensate for the
inevitable rise in pressure observed with increased distance from
the source of vacuum. Before discussing specific options for the
forepump and diffusion pump, several general design considerations
should be noted which serve to minimize the magnitude of this pressure rise.(1)

To reduce conductance losses in pumping, all connections
linking forepump and diffusion pump should be as short as possible.
Connecting lines should incorporate a minimum number of bends, and
those that must be used should be gradual curves rather than sharp
angles. Finally, wherever economically practical, connections
should be made with large (we use 1") diameter heavy wall stainless
steel flexible tubing. Cajon Flexible Tubing equipped with Cajon
Ultra-Torr unions and adapters (2) is adequate for this purpose
although it must be telescoped down to its minimum flexible length
before vacuum is established to prevent collapse of the flexible
section once vacuum is applied. A better choice are the heavier
gauge stainless steel flexible tubings available from a number of
manufacturers.(3,4) This tubing is available in a variety of
lengths and comes fitted with standard ISO 25 mm flanges for direct
connection to metal vacuum pump couplings. The least desirable
choice is heavy wall rubber vacuum tubing which is susceptible to
cracking and prolonged outgassing under continued laboratory use.

FOREPUMPS

General forepump requirements are quite simple. The pump should be
capable of attaining a minimum ultimate vacuum of 10^{-3} torr with a
minimum pumping speed of 60 1/min. Higher pumping speeds are
preferred if practical. In addition, the pump should be as quiet,
rugged and chemically resistant as possible. Both belt drive and
direct drive pumps are capable of satisfying these requirements.
Belt drive pumps are extremely rugged and tolerate pressure surges
quite well. However, they can be noisy and cumbersome (the higher
the pumping speed, the larger the pump). Direct drive pumps are
normally very quiet and quite small for their pumping speed.
However, they are not as chemically resistant as belt-driven rotary
pumps (although a few highly chemically resistant versions are
available) and cannot tolerate sustained pressure surges without
overheating.

In our laboratory, both types of pumps are in operation on high
vacuum systems. For high speed pumping stations, the Edwards Model
E2M18 direct drive pump with a pumping speed of 415 1/min is used.
Smaller vacuum lines and hybrid Schlenk lines are pumped by small
capacity (60-160 1/min) Welch Duo-Seal models.(5) Since both pump
types are suitable for high vacuum pumping stations, the choice of
pump will depend on the specific advantages and disadvantages given
special weight by the individual researcher.

DIFFUSION PUMPS

Diffusion pumps employed in high vacuum pumping stations should contain pumping fluids that are relatively inexpensive and are oxidatively and chemically inert. They must be capable of attaining ultimate vacuums of $10^{-6} - 10^{-7}$ torr and should have the highest pumping speeds that are economically practical.

Two types of diffusion pumps are commonly employed in high vacuum pumping stations. They are the mercury diffusion pump and the oil diffusion pump. The latter can be constructed of either metal or glass. The mercury pump is usually constructed only of glass. Although mercury pumps are common in chemical laboratories due to their ease of construction and long history of use, we strongly advise against their use for the following reasons. Mercury pumps must be trapped to avoid diffusion of mercury vapor into the working vacuum line. They must be continuously water cooled, with all the attendant precautions this implies. In addition, mercury is an insidious environmental poison which requires extreme caution in using and venting the pump. Finally, oil pumps are available which are comparably priced and give superior performance. These pumps use fluids (6) which have very low vapor pressures ($10^{-7} - 10^{-10}$ torr) at their working temperatures, making contamination of the vacuum line by the pumping fluid rare. (In practice, however, they are commonly trapped to protect the hot oil from chemical contamination or oxidation). In addition, most glass oil pumps and some small metal oil pumps are air-cooled, thereby avoiding the need for continuous water circulation through the pump body.

If price is not a major concern, metal oil diffusion pumps are preferred. They are available in virtually any size allowing extremely high pumping speeds to be obtained. They are quite rugged and durable and easily cleaned. If an accident should occur and the pumping fluid is disastrously oxidized, a metal pump body can always be cleaned by sand-blasting or liquid honing. Finally, unlike glass pumps, metal pumps are not easily broken.

We have found two metal pump designs to be of great practical utility in our laboratory. Small lines are conveniently pumped by the relatively inexpensive air-cooled VEECO 3-in. throat pump (120 l/sec pumping speed). High capacity, high speed pumping stations incorporate the water-cooled Edwards 100M Diffstak (322 l/sec pumping speed).(7) Both pumps have proven extremely reliable in long-term operation.

BYPASS DESIGNS

Consideration must next be given to the type of bypass design used in the system. In going from high vacuum to low vacuum (such as when an apparatus is first evacuated on the line), some mechanism must exist for isolating the diffusion pump from the pumping station while the forepump roughpumps the system. If this is not done, the operation of the diffusion pump will be momentarily halted and the hot pumping fluid will be degraded by chemical and oxidative processes. The most common type of bypass is the three valve design illustrated in Figure 1. In bypass mode, valves A and B are closed,

Figure 1.   3-Valve bypass design used on medium vacuum/Schlenk lines.
Note position of valves labelled A, B and C.   Also note 1.)
the compact VEECO oil diffusion pump and 2.) the grease-free
57 mm O-ring flat-flanged trap used on this type of line.

isolating the diffusion pump. Valve C is then opened allowing the
forepump to pump the line directly. When the system pressure has
dropped to 10 torr or less (see next section), the procedure is
reversed and the diffusion pump is restored to the circuit. This
type of valving system is relatively simple and inexpensive. However, it is cumbersome in operation since three valves must be
manipulated to effect the bypass. Although we use such bypasses on
our small vacuum lines and hybrid Schlenk lines, we much prefer the
butterfly valve/3-way valve design sold by Edwards in conjunction
with their Diffstak line of pumps (Figure 2,2a). In this configuration, a butterfly valve is fitted to the throat of the diffusion
pump body through a spacer collar flange assembly. The 3-way valve
can be operated to provide backing of the diffusion pump, direct
access to the forepump or a closed neutral position. In practice,
it is a simple matter to isolate the diffusion pump by throwing the
butterfly valve closed and setting the 3-way valve to roughing.
Reversing the procedure restores the Diffstak to line pumping.
Although such an assembly is expensive it should be considered the
preferred choice for bypass designs. If this assembly cannot be
used, careful thought should be given to the layout and ease of
operation of the 3 valve bypass design. Experience indicates that
well-designed bypasses will be used, while inconvenient ones will
not; to the detriment of the performance and useful lifetime of the
pumping station.

LINE CONNECTIONS

The type of connection used to attach the pumping station to the
working vacuum line will be dictated by the type of diffusion pump
used and the material of its construction. If a metal pump is used,
connections can be made directly to standard ISO 25 mm flanges using
either rigid connectors such as tees or bends or through the use of
heavy gauge ISO flanged flexible tubing. Alternately, Cajon Ultra-
Torr connectors can be used on straight lengths of 1" diameter tubing. If attachments to glass must be made, Cajon fittings correctly
selected for diameter and length are the connectors of choice.
Again, care must be taken not to clamp Cajon flexible tubing rigidly
before introducing vacuum in the system.(4)
     Many potential variations exist on these simple themes. For
example, the high capacity pumping station used in our laboratory
can be connected to any of five different vacuum lines by attaching
standard flanges to an octahedral cross welded to the face plate
flange bolted to the diffusion pump. Similar variations are possible with glass systems fitted with the appropriate hardware.

VACUUM GAUGES

Reliable, rugged and chemically resistant vacuum gauges must be
properly installed (optimally, away from cold traps, which will give
spuriously low gauge readings) in the pumping station and in the
working sections of the lines. It is advantageous if these gauges
read from atmospheric pressure to the high vacuum range. However,
in practice, no one gauge is capable of indicating reliably over
such a wide pressure range, and at least two separate gauges are

Figure 2.   Cut-away view of the Diffstak pump and bypass system.
The butterfly valve (A) is shown in the closed position with the
spacer collar assembly eliminated for clarity.   Continued on next page.

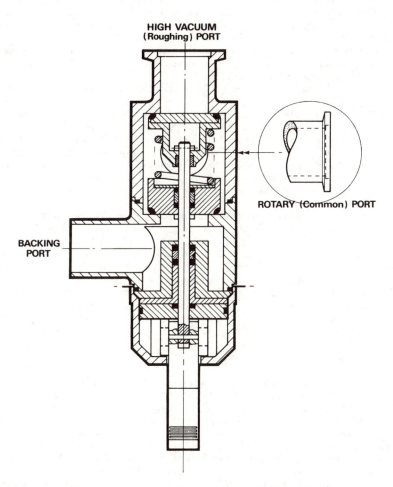

HIGH VACUUM
(Roughing) PORT

ROTARY (Common) PORT

BACKING
PORT

Figure 2.   Continued.   An interior view of the backing/roughing valve.   (Illustration kindly provided by Edwards High Vacuum, Inc.)

normally used in concert to cover the working range of the line. We will consider three pressure ranges which are of interest to synthetic organometallic chemists: 1-1000 torr, $1-10^{-3}$ torr and $10^{-3} - 10^{-7}$ torr.

The first range can be adequately handled with a simple mercury manometer.(1) However, absolute pressure transducers, such as the Baratron 200 Series manufactured by MKS Instruments, offer much greater accuracy over a wider pressure range.(8) If accurate and precise pressure measurements are commonly conducted on the vacuum line, the transducers are the sensors of choice. If not, mercury manometers are the economical alternative.

The pressure range $1 - 10^{-3}$ torr can be handled by either a McLeod gauge(1) or a simple thermocouple gauge (such as the Varian 801).(9) Since McLeod gauges are often difficult to read and cumbersome to operate, the inexpensive, rugged and easy-to-read thermocouple gauges available from a number of manufacturers are the best choice in this pressure regime. Thermocouple gauges should be installed wherever needed in the vacuum line systems. As noted in the previous section, one should be installed at the pumping station to provide a means to gauge when the diffusion pump bypass is necessary.

Rugged, chemically resistant, high vacuum gauges are difficult to find. In our experience, only the cold-cathode ionization gauges can reliably fulfill these requirements.(10) Optimally, several such gauges should be installed in the vacuum system. One gauge must be located at the pumping station to record the performance of the diffusion pump. Other gauges should be located in the working sections of the line to provide accurate pressure readings away from the pumping station. Losses in pumping efficiency due to conductance can sometimes result in pressures which are at least one order of magnitude greater than they are at the pumping station.

VACUUM LINES

Vacuum line designs for the synthetic organometallic laboratory fall into two general categories: single manifold designs intended solely for high vacuum work, and double manifold designs constructed for the manipulation of materials under conditions of high vacuum or inert atmosphere including the introduction or removal of solvent. Since most design features are independent of the number of manifolds employed, we discuss these systems together. In a separate section dealing with specialized vacuum line designs, we consider the utility of specific double manifold designs for Schlenk-type work.

Any synthetic vacuum line should incorporate the same basic design features already addressed in the discussion of pumping station design (vide supra). In addition, all vacuum lines must be designed so that they can easily be trapped to protect the diffusion pump oil from degradation and to prevent the venting of hazardous vapors into the laboratory. Although conventional standard taper ground glass joint traps are adequate for this purpose, we favor greaseless traps constructed with 57 mm flat O-ring flanges.(11) These traps are reliable high vacuum components which, unlike greased traps, will not freeze after prolonged use. Trap-to-line and trap-to-pumping station connections are made with 1/2" Cajon Ultra-Torr unions and the minimum length of flexible stainless steel tubing necessary to properly align components. In normal use, one trap per line is adequate. Double-trapping may be necessary if extremely volatile materials are being manipulated.

Specific design considerations focus on the type of valve or stopcock used in the line and the type of connector used to attach apparatus to the line. Although vacuum lines have been traditionally constructed using greased high vacuum stopcocks and greased connectors such as ball joints or standard taper joints, modern advances in Teflon valve construction and elastomer chemical resistance have provided an excellent alternative. In our laboratory, both single and double manifold lines are constructed of 8 mm Kontes hi-vacuum Teflon valves (12) (one exception to this is the Schlenk line discussed in a later section). Elastomers compatible with the solvents used in the system are chosen for the valve seats (backing O-rings are protected by Teflon wipers in the Kontes tap design; the choice of O-rings here is not critical). Kalrez, a DuPont product, which is resistant to most common solvents including THF, is the seat O-ring elastomer of choice, but is quite expensive.(13) Viton is a good general purpose substitute. Special use elastomers (for example, the use of ethylene-propylene copolymer O-rings for work with liquid ammonia) can easily be chosen by consulting the elastomer resistance charts provided by several O-ring suppliers. It should be noted that when first assembled, a vacuum line based on O-ring valves (which trap gases between the backing O-ring seats) must be pumped for some time before a good vacuum is obtained.

Connections to the line are made through 15 mm O-ring joints or with 3/8" Cajon Ultra-Torr fittings.(14) Conveniently, most Kontes Schlenkware will mate with 3/8" Cajon connectors. O-ring glassware for use with the lines is unfortunately not yet commercially available. However, any chemical glass company will normally tool such items as requested by affixing O-ring joints to glassblower blank flasks. The price of the glassware is not prohibitive, normally no more than the price of the two items combined. The most useful flasks and connectors which we have constructed for use with these lines are illustrated in Figure 3. A typical assembly (shown in Figure 4) depicts a vacuum transfer apparatus and illustrates the convenience and ease of assembly and disassembly typical of this type of modular line and equipment design.

By incorporating these various general and specific design features, we have constructed two simple vacuum lines, a single and a double manifold design, which are in everyday use in our laboratory. The double manifold is shown in Figure 5.(15) The basic design can be easily modified to lengthen or shorten the line or to accomodate fewer or more ports. The manifolds are also scaled such that the entire vacuum line system can be contained in a single fume hood. One further design feature deserves comment. On the single manifold system (not shown), it is desirable to mount the valves at a 45° angle to the body of the line. This simple modification allows a clear side-on view of the seat made by the valve and reduces the possibility of overtightening and snapping the valve.

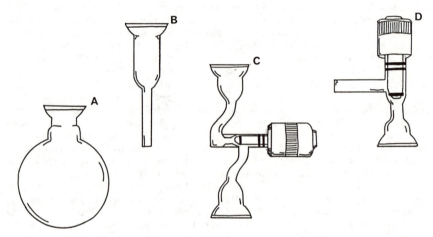

Figure 3.   Commonly used glassware and connectors.   A.) 100 mL RB flask
            equipped with 15 mm O-ring joint.   B.) 15 mm O-ring to 3/8"
            medium wall tubing adapter.   C.) 15 mm O-ring Teflon valve
            adapter (both 4 and 8 mm sizes are useful).   D.) 15 mm O-ring
            to 3/8" medium wall tubing Teflon valved adapter.

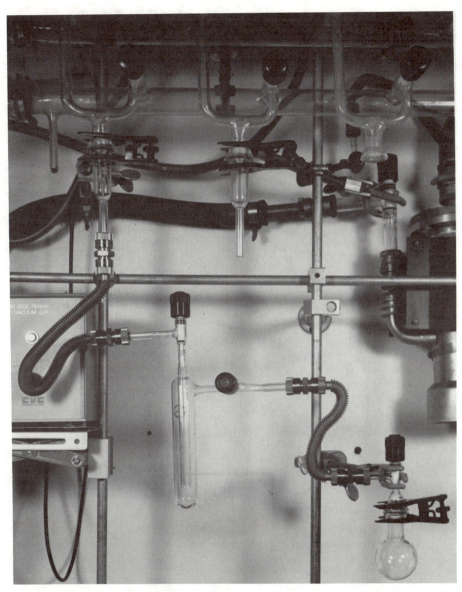

Figure 4.   Flex tubing connected vacuum transfer apparatus.
<u>Continued on next page</u>.

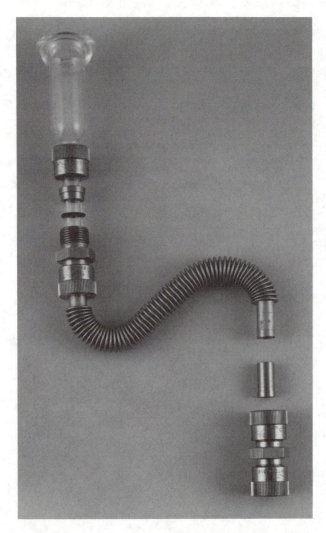

Figure 4. Continued. An exploded view of the Cajon flexible
tubing and unions used in this assembly.

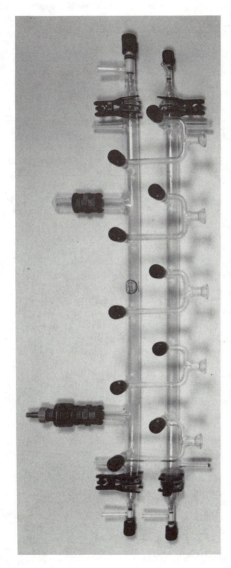

Figure 5. Standard 5-port Teflon valved double manifold vacuum line. The line incorporates 8 mm Teflon valves and 20 mm O-ring end caps individually valved with 4 and 8 mm 90° Teflon valve adapters. Gauge tubulations are 1" in diameter. A spare port is included to provide independent checking of gauge performance by installing a second calibrated gauge tube.

SPECIALTY DESIGNS

In addition to employing the double manifold line for high vacuum work, we have also used the manifold as the basis for hybrid high vacuum/Schlenk and NH_3 lines. In the former configuration, the lower manifold functions as an over-pressure inert atmosphere line modeled after the basic Ace-Burlitch gas handling system. The general layout of this system is the same as that of the medium vacuum/Schlenk line (vide infra). The upper manifold is normally operated at roughing pressures by isolating the diffusion pump from the system. However, if a high vacuum application arises, the pump is readily brought into the circuit by eliminating the bypass. This type of system is quite useful when high vacuum work is only occasionally necessary. The working layout of the high vacuum/NH_3 line is illustrated in Figure 6. Its detailed design and synthetic applications have been described previously.(16)

For routine Schlenk-line work where access to high vacuum is unnecessary, the diffusion pump is eliminated and the line is pumped directly by the roughing pump (with appropriate trapping). For the working line, we favor a double manifold design using greased double oblique bore solid stopcocks as illustrated in Figure 7. This system allows rapid cycling between rough vacuum and inert atmosphere manifolds with one smooth turn of the stopcock; note that in the greaseless valved design, two valves must be manipulated to effect the same result. As in the high vacuum/Schlenk line design, water and oxygen impurities are removed from the inert gas by its passing first over a column of MnO supported on Vermiculite and then over a column containing a drying agent ($CaSO_4$/activated 4A molecular sieves). The MnO column is placed so that it can easily be wrapped with heating tape when the catalyst needs regeneration (accomplished by heating to 330°C while passing pure hydrogen slowly over the support). The rough vacuum line can be fitted with one or two cold traps for removal of solvents from reaction mixtures; if a mercury manometer is included in the vacuum line for pressure measurements, it should be located between the two traps to prevent contamination of the pump by mercury vapor. Flat 57 mm O-ring flanges equipped with greaseless Teflon valves are used as connectors on the ends of the drying and catalyst columns; O-ring joints are used on the ends of the double manifold. These connections facilitate replacement of the drying and deoxygenating reagents, removal and cleaning of the manifold, or isolation of one component of the system. All connections between components of the inert gas train and vacuum line are made with Cajon Ultra-Torr connectors and flexible stainless steel tubing; the line outlet ports can be connected to apparatus by similar Cajon connectors and flexible stainless steel tubing or by heavy-walled rubber vacuum tubing.

We have also constructed and used lines based on the Louwers Hapert two way high vacuum valves described by Landis (see succeeding article). Although we were initially quite pleased with the design of this valve (for the same reasons outlined by Landis'), we abandoned (our systems based upon) it when one aspiration of THF into the line irreparably froze the taps in place as the THF swelled the Viton O-rings. Hence, we agree with Landis' recommendation that these taps be refitted with Kalrez O-rings (17) to avoid such a danger.

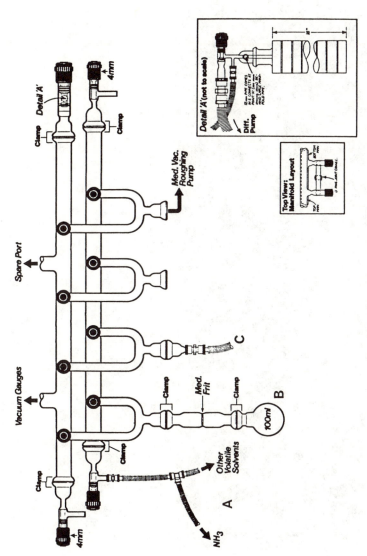

Figure 6. Working layout for a hybrid high vacuum/NH₃ handling line. Note A.) NH₃/volatile solvent inlet system. B.) assembly for pre-drying ammonia. C.) connection to reaction apparatus. Figure reproduced from Ref. 16a. Copyright 1985, American Chemical Society. Further details are available from the publications cited in reference 16.

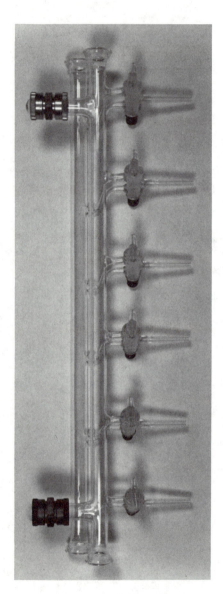

Figure 7. Greased, double oblique bore stopcock, double manifold medium vacuum/ Schlenk line.

SUMMARY

We have described several vacuum line systems which we have successfully used in our laboratory for a wide variety of manipulations and
syntheses. Although the initial costs of some of the equipment and
materials used may seem prohibitive, in our experience the added
versatility and ruggedness of design that these systems provide is
well worth the initial investment.

ACKNOWLEDGMENTS

ALW would like to thank her colleagues at AT&T Bell Laboratories for
their many useful suggestions regarding these vacuum line designs as
they have evolved over the years. She would also like to acknowledge Professor Philip E. Eaton's role in introducing her to the
aesthetic delight associated with using a well-designed piece of
laboratory equipment. She is also grateful for the opportunity to
work with Professor James L. Dye who taught her the important lesson
that enthusiasm for life and science is not correlated with age.

REFERENCES AND NOTES

1. An excellent discussion of vacuum line designs and related
   topics may be found in "Experiments in Physical Chemistry," D.
   P. Shoemaker, C. W. Garland, and J. I. Steinfeld. McGraw-Hill:
   New York. 1974 pp. 579-625 and in "The Manipulation of Air-
   Sensitive Compounds," 2nd Edition, D. F. Shriver and M. A.
   Drezdon. John Wiley and Sons: New York. 1986.

2. Cajon Stainless Steel Flexible Tubing is available through R. S
   Crum and Co., 1181 Globe Avenue, Mountainside, New Jersey
   07092 (201-232-4444). Other local suppliers may also handle
   the line. 1-in. o.d. 12-in. flexible length tubing, 321-16-X-
   12, $80.40. 1-in brass union, B-16-UT-6, $23.20 (also available
   in stainless steel). Sleeve inserts for 1-in. o.d., 304-16-
   XOA, $4.00 each.

3. Leybold-Heraeus Inc., 5700 Mellon Road, Export, PA 15632 (412-
   327-5700). 10-in. flexible stainless steel hose fitted with 25
   mm (1") KF ISO standard fittings, 86783, $58.85. 20-in. flexi-
   ble length, 86793, $87.75. Mating fittings are available from
   Leybold-Heraeus or from MDC, 23842 Cabot Boulevard, Hayward,
   California 94545 (415-887-6100). Centering ring assemblies and
   KF clamps for connections must be ordered separately. Representative fittings and prices are given for MDC products. 1-
   in. Kwik-Flange (KF) aluminum clamp, K 100-C, $7.00. 1-in.
   KF half-nipple, K 100-1, $15.00. 1-in. KF elbow, K 100-2L,
   $43.00. 1-in. KF tee, K 100-3, $70.00. 1-in. KF 6-way cross, K
   100-6, $190.00.

4. When first installed in a vacuum system, flexible tubing
   requires substantial pumping to completely desorb moisture and
   other contaminants from its high surface area. Desorption may
   be speeded by wrapping the piece with heating tape and baking

the section until a reasonable vacuum is established. In addition, glass apparatus connected by flexible tubing should not be   **rigidly**   clamped until full vacuum is established in the system. Large-bore flexible tubing will contract when evacuated and must be allowed to reach its final evacuated length before glass components are firmly clamped.

5. Edwards High Vacuum Inc., 3279 Grand Island Blvd., Grand Island, NY 14072 (716-773-7552). E2M18 Direct Drive Pump, $1450.00. Sargent-Welch Scientific Co., 35 Stern Avenue, Springfield, NJ 07081 (201-376-7050). 1405B-01 2-Stage, 60 l/m rotary pump, $1075.00. 1402B-01 2-Stage 160 l/m rotary pump, $1205.00.

6. An economical diffusion pump oil is Dow-Corning Silicon Oil 705 available from Edwards ($162.00/500 mL). If price is not a concern, the premium oil is Santovac 5 ($433.00/500 mL) also available from Edwards.

7. VEECO Instrument Co., Terminal Drive, Plainview, NY 11803 (516-349-8300). 3" diameter air-cooled diffusion pump, 4530-009-02, $1855.00. Available from Edwards High Vacuum Inc., Edwards Diffstak MkII, Model 100M, 050B346-31-000, $1990.00 equipped with the BRV 25K backing/roughing valve, 08-C323-03-000, $325.00. All connecting hardware and blank flanges for system connections must be ordered separately. In addition, the Diffstak requires installation of a spacer collar assembly if the unit is to be used with a simple capping flange.

8. MKS Instruments Inc., 6 Shattuck Road, Andover, MA 01810 (617-975-2350). 122AA-01000 1000 mm Pressure Transducer (other ranges and sensitivities are also available), $645.00 and the PDR-5B Power Supply/Readout, $995.00 (5-port sensor; smaller, less expensive units are also available).

9. Varian Associates, 331 Montvale Avenue, Woburn, MA 01801 (617-935-5185). 0801-F2739-301, Model 801 Thermocouple Gauge, $186.00 complete. Replacement sensing tubes are also available.

10. CVC Products, P.O. Box 1896, Rochester, NY 14603 (716-458-2550). GPH-320C Vacuum Penning Gauge with Sensor (Part #280460), $998.00 complete.

11. Lab Glass Inc., P.O. Box 610, NW Blvd., Vineland, NJ 08360 (609-691-3200). LG-1024-102 O-ring joint, flat, size 57 mm, $22.90. O-rings and clamps are ordered separately.

12. Kontes, Spruce Street, P.O. Box 729, Vineland, NJ 08360 (609-692-8500). Kontes K-826510-0004: 0-4 mm, $33.20. K-826510-0008, 0-8 mm, $53.00. K-826510-0012, 0-12 mm, $61.70. We have also used 5, 10 and 15 mm ace valves on our single manifold designs without incident. Ace Glass Inc., 1430 NW Blvd., P.O. Box 688, Vineland, NJ 08360.

13. Kalrez O-rings are available from Service Seal and Packing, Inc., 277 Fairfield Road, Fairfield, NJ 07006 (201-575-4277) Size 011 (Kontes 8 mm valve seat size), $15.80. Size 116 (15 mm O-ring joint size), $27.90.

14. Cajon Ultra-Torr 3/8" brass union, B-6-UT-6, $8.00. In stainless steel, SS-6-UT-6, $17.20. Flexible stainless steel tubing: 3/8 in. o.d., 24-in. flexible length, 321-6-X-24, $52.20. Sleeve inserts for 3/8 in. o.d. 304-6-X0A, $2.50 each.

15. The complete manifold (including vacuum trap) is available from Crown Glass Co., 990 Evergreen Drive, Somerville, NJ 08876. Quotes furnished upon request.

16. a.) Wayda, A. L.; Dye, J. L. J. Chem. Ed., 62, 356 (1985) and b.) Wayda, A. L.; Dye, J. L.; Rogers, R. D. Organometallics, 3, 1605 (1984).

17. In our experience, Kalrez O-rings do not have the same mechanical resiliency as Viton. Therefore, all Kalrez O-rings should be periodically checked to assure that the size and structural integrity of the O-ring remains intact.

RECEIVED July 31, 1987

Chapter 5: Application 1

Giving Labs the Dry Look: Greaseless Double Manifolds and Schlenkware

Clark Landis

Department of Chemistry and Biochemistry, University of Colorado, Boulder, CO 80309-0215

Manipulations of air- and water-sensitive compounds are accomplished easily using double manifolds and Schlenk-type glassware, such as the Ace-Burlitch Inert Atmosphere system and No-Air Labware. While these systems have proven performance and reliability, the advent of new materials and devices allows improvement of the basic Ace-Burlitch system to a system which is completely greaseless and truly high vacuum. The greaseless double manifold (Figure 1) is based on the Louwers Hapert two-way high vacuum stopcock (12 mm bore, Aldrich #Z11,293-3). The unique (to my knowledge) configuration of this two-way high vacuum stopcock enables one, with ca. four quick turns of the wrist, to switch between a high vacuum manifold and a pressurized gas manifold. The "O" ring seals on the valve plunger assure true high vacuum (10^{-6} mm Hg) performance, allow super-ambient (15 psig) gas manifold pressures, and make for a low-maintenance, easily cleaned double-manifold. My double manifold has the two-way valves angled at 45° to make the "O" ring seals (your indication of the valve position) easily visible. Also used are Ace high-vacuum "O" ring valves (Ace #8194) which connect (1) the double manifold to the purified gas supply, (2) the vacuum manifold to the solvent trap, and (3) the vacuum manifold to a thermocouple gauge. One end of each of the manifolds is equipped with a "O" ring sealed Teflon plug screwed into Ace-threads (Ace #5846 and #7644) to give access to a bottle brush for cleaning the manifolds. The two manifolds, via the two-way valves, connect to an assembly comprising a 14/35 "O" ring joint (Ace #7648) and a 3 mm high vacuum valve (Ace #8192) which connects to a mercury bubbler. Schlenk-type glassware is connected to the system either by direct connection to the #14/35 "O" ring inner joint or by connection via butyl rubber tubing. Note that the double-manifold does contain one vestigial greased connection - the ball-and-socket joint connecting the vacuum manifold to the trap is greased. I have found that the "O" ring ball joints do not make good high vacuum seals with a ground glass socket, however, the stress reduction afforded by a flexible ball-and-socket joint reduces breakage during assembly and disassembly.

0097-6156/87/0357-0136$06.00/0

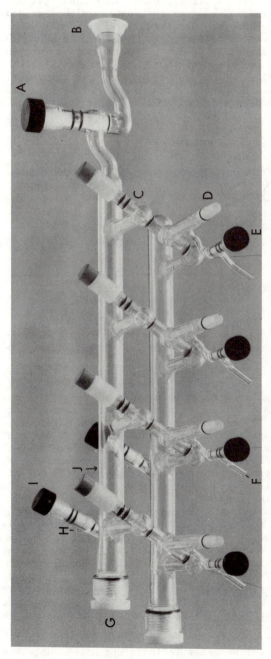

Figure 1. Greaseless double manifold. A) 12 mm. "O" ring valve. B) Ball-and-socket connection to trap. C) Louwers Hapert 12 mm. two-way stopcock D) #14/35 "O" ring inner joint E) 3 mm. "O" ring valve F) Connection (via butyl rubber tubing) to mercury bubbler; all four valves share a common bubbler G) Threaded glass joint with teflon plug and "O" ring seal H) Connection (via Cajon union) to thermocouple guage I) 5 mm. "O" ring valve J) Connection (via bellows tubing and Cajon union) to inert gas supply. (Gene Lutter, University of Colorado, glassblower)

A particular problem in using ground glass inner joints containing a groove and "0" ring is that mating with a ground glass outer joint rapidly wears the relatively soft elastomer with concomitant loss of vacuum performance. However, this problem is easily overcome by the use of Clear-seal outer joints (Wheaton #758966). The smooth surfaces of these joints make excellent high vacuum seals with "0" rings without destroying the elastomer. I have designed and used completely greaseless Schlenk-type glassware (see Figure 2) by incorporating Clear-seal outer joints, "0" ring

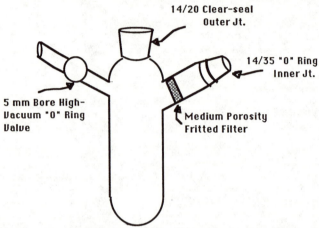

14/20 Clear-seal Outer Jt.

14/35 "0" Ring Inner Jt.

5 mm Bore High-Vacuum "0" Ring Valve

Medium Porosity Fritted Filter

Figure 2. Greaseless Schlenk-type reaction/filtration flask.

inner joints and 5 mm high vacuum Teflon valves into the traditional Schlenk-type designs. The resulting glassware features the followng attributes: (1) the glassware can be used on either a high vacuum line or on a double-manifold, (2) grease is not a contaminant, (3) the tight fit of the "0" ring inner joint with the Clear-seal outer joint accommodates super-ambient pressure (useful for cannula techniques, difficult filtrations, etc.), (4) the length of the #14/35 "0" ring inner joints prevents solutions and solids from coming into direct contact with the "0" rings, and (5) the glassware is completely compatible with other glassware utilizing standard taper joints (including greased).

A persistent problem with the use of elastomeric "0" rings is swelling by solvents such as THF. Substitution of Viton by Kalrez solves the problem but is very expensive and I have found the Kalrez "0" rings to be more susceptible to tearing. (N.B. long term exposure of the Louwers Hapert two-way valves may swell the "0" rings and cause the valve stem to "freeze" in place. If long term exposure to THF is anticipated, Kalrez "0" rings should be used.) The lifetime of the "0" rings in the glassware described above can be lengthened considerably by using Teflon sleeves on the inner joint (Teflon tape applied directly to the "0" ring may help), by lubricating the "0" rings with Krytox (Aldrich #Z12,307-2) fluorocarbon grease, and by using Teflon plugs with double Teflon ring seals (Ace #8192B) in the high vacuum valves.

RECEIVED June 12, 1987

Chapter 6

Systems That Allow Diverse Synthesis and Purification Procedures Within an Inert Atmosphere Glovebox

T. W. Weidman

AT&T Bell Laboratories, Murray Hill, NJ 07974

While inert atmosphere gloveboxes greatly facilitate the manipulation and storage of air sensitive materials, standard systems are often unsuitable for synthesis and purification procedures. A variety of apparatus and techniques have been developed to allow otherwise complex or hazardous procedures to be performed entirely inside a glovebox. These include a recirculating inert liquid heat exchange system to provide localized cooling for high temperature processes and a refrigerated immersion probe feedthrough for low temperature operations and for the condensation and removal of solvent vapors. In addition, a compact system for preparative liquid chromatography is described, together with feedthroughs and procedures for direct anaerobic transfer of liquids into and out of a glovebox.

Inert atmosphere glovebox systems are extensively employed in numerous fields of research requiring the manipulation of air and moisture sensitive materials. Commercially available units allow the convenient measurement, transfer and storage of reagents and are often used to perform simple concentration, filtration, and crystallization procedures. However air sensitive reactions requiring elevated and/or sub-ambient temperatures or large volumes of solvents have traditionally been conducted externally, with apparatus specially designed to exclude oxygen and moisture.

In preparation for research involving the synthesis of zero-valent transition metal complexes, new apparatus and techniques have been developed which allow otherwise complex and tedious procedures to be performed entirely within a glovebox. The following discussion will focus on the assembly and application of those new systems which provide capabilities potentially useful in many areas of research. No attempt will be made to review standard glovebox procedures or commercially available equipment, on which a good introduction is already available.(1)

0097–6156/87/0357–0139$06.00/0
© 1987 American Chemical Society

The apparatus described herein has been installed and operated
in perhaps the most familiar basic glovebox, the Vacuum/Atmospheres
Co. (VAC) Model HE-43-2.(2) However, it should be no less feasible
to assemble the same or similar equipment into any other rigid wall
inert atmosphere enclosure. Most of the systems described here
contain components which are available from a number of manufacturers and distributors. In the interest of clarity, and to provide
ready access to additional information, a supplier is occasionally
identified. Such information does not necessarily imply a recommendation, and the interested reader is encouraged to consider alter-
nate sources and make current cost comparisons. Finally, there is
no intent to suggest that the use of these systems is any more or
less safe than conventional procedures outside of a glovebox, and as
always the chemist must evaluate each alternative (when planning a
specific reaction or procedure) on a case by case basis.

Design and Assembly Descriptions

A Recirculating Inert Liquid Heat Removal System. While standard
inert atmosphere gloveboxes may be used to perform ambient temperature reactions, the lack of a facility for localized heat removal
(as provided externally be running tap water) precludes all syntheses requiring reflux, distillation, or sublimation. This limitation
may be eliminated by the installation of a recirculating liquid heat
exchange system.

A basic assembly (illustrated in Figure 1) employs a small
adjustable flow pump to circulate a non-reactive solvent between
standard glass cooling apparatus (condensers, sublimators, etc.) and
an externally cooled 1/4 inch copper coil. The potential for leakage of water or oxygen into the coolant (and potentially into the
glovebox) is minimized by forming the only external component, the
heat exchange coil, from one continuous length of copper tube, both
ends of which extend into the glovebox. A feedthrough configuration
(see Figure 2) is employed which routes the 1/4" OD tube straight
through the interior of a standard 1/2 FPT glovebox service port.
The fittings indicated provide a double seal against the copper tube
while otherwise minimizing close contact and heat transfer. Additional insulation is provided by injecting foam-in-place polyurethane sealant into the feedthrough interior before final assembly.

External heat removal may be accomplished with chilled flowing
water, as provided by a building heat exchange system or (for temporary applications) with a simple ice bath. Greater versatility
and temperature control may be achieved using a variable temperature
refrigeration bath, but care must be taken to retighten all feedthrough fittings at a minimum operating temperature.(4)

The circulation pump should produce flow rates up to at least
500 ml/minute and provide continuous and consistent output over the
anticipated temperature range (ca. -20° to 40°C). In addition all
"wetted" pump materials must exhibit long term compatability with
the liquid selected, or preferably (for maximum versatility) any
organic solvent. These requirements generally favor a positive
displacement design, as opposed to diaphragm, bellows, or centrifu-
gal action pumps. Satisfactory performance has been obtained from a
laboratory metering pump (manufactured by Fluid Metering, Inc).(4)
incorporating a stainless steel pump head, an alumina ceramic piston
and liner, and Teflon (5) seals.

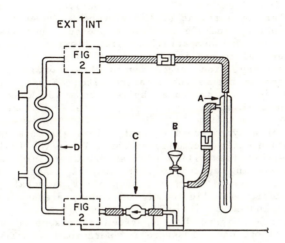

Figure 1   Recirculating Inert Liquid Removal System. A: standard glass cooling apparatus. B: liquid reservoir, C: adjustable pump, D: heat exchange coil.

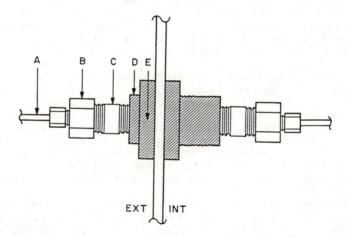

Figure 2   Heat Exchange Tube Feedthrough. A: 1/4" OD copper tubing, B: "bored through" Swagelock 1/2 FPT to 1/4 tube adaptor(3), C: 2"-1/2-MPT nipple, D: standard VAC 1/2 FPT service port, E: feedthrough interior insulated foam-in-place polyurethane.

The selection of an "inert" or "non-reactive" liquid may depend on the particular applications of the glovebox. However, desirable properties ordinarily include low vapor pressure and viscosity, a broad liquid range, and compatibility with flexible tubing materials. For organometallic applications an additional consideration is poor ligating ability, since trace levels of a coordinating coolant might compete for an active coordination site. These criteria may be met by a variety of saturated hydrocarbons (such as decane or decahydronaphthalene) which can be prepurified by distillation at reduced pressure from sodium.

Selection of a hydrocarbon coolant allows all internal connections to be conveniently accomplished using Viton (5) tubing, which retains considerable flexibility even at -20°C. Rapid addition or interchange of cooling apparatus with minimal leakage is further facilitated by the use of automatic shut-off type metal quick disconnects.(6) Additional liquid may be added, pressure equalized, and bubbles eliminated by including a reservoir bottle similar to that illustrated in Figure 1. A system so assembled allows the convenient use of ordinary glassware in operations normally performed with water cooling.

Low Temperature Apparatus. Numerous synthetic procedures requiring low temperatures are routinely performed (outside of a glovebox) utilizing a dry ice/solvent slush bath for temperature control. One system allowing internal operations at low temperatures is available as a factory installed option on the VAC gloveboxes.(2) This accessory is essentially a bucket extending through the glovebox floor, which may be surrounded externally with a suitable cold bath. The unit occupies no internal space, and is best suited for short term cryogenic storage.

Alternative apparatus providing convenient maintenance of low temperatures is illustrated in Figure 3. The basic assembly consists of an externally located refrigeration compressor unit connected via an insulated hose to a flexible cooling probe inside the glovebox. Refrigeration units are available which provide continuous cooling and/or control of temperatures in the range of - 20°C to - 100°C. The greatest versatility in internal cooling configuration may be achieved with units offering the longest and most flexible immersion coils, such as Neslab Inc. Model CC-60, CC-80 or CC-100 with "Type FV" probes.(7)

Regardless of the unit selected, a sturdy, leak free, seal into a glovebox feedthrough is best accomplished by requesting than an insulating 1" OD sheath be attached to the probe during initial factory assembly. The sheath or collar should be welded over the smooth tube between the insulated hose and the flexible probe and be either evacuable or permanently evacuated (see Figure 4). With this modifiction the probe may be easily installed (and if necessary removed) using a 1" FPT service port (the largest standard VAC feedthrough) and a bored through Swagelock(3) (exterior side) and bored through Ultratorr(8) (interior) 1" tube adapter.

Once the flexible probe has been installed, a versatile cold bath configuration is obtained by coiling the probe (while at ambient temperature) to fit a low form Dewar flask or equivalent. Positioning the coiled probe approximately one foot from the glovebox

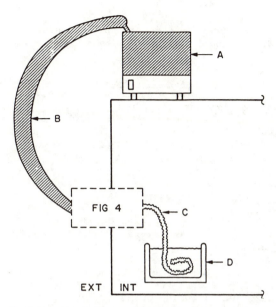

Figure 3   Refrigerated Immersion Probe Cold Bath Assembly.
A: external compressor unit, B: insulated hose,
C: flexible bellows immersion probe, D: low form Dewar
flask for cold bath.

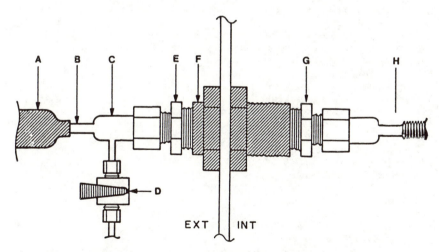

Figure 4   Refrigerated, Immersion Probe Feedthrough Assembly.
A: insulated hose, B: 3/8" OD straight tubing, C: 1" OD
evacuable sheath, D: evacuation valve, E: 1" OD "bored
through" Swagelock tube to 1" MPT adapter, F: 1" FPT
VAC service port, G: 1" OD "bored through" Cajon
Ultratorr(8) tube to 1" MPT adapter, H: flexible
bellows immersion probe.

floor permits the use of a magnetic stirrer or laboratory jack under
the cold bath and allows the probe to be suspended above the bath
(or other collector), facilitating its use for the removal of stray
solvent vapors.

Liquid Transfer and Chromatography Techniques. A common drawback to
performing synthetic operations inside a glovebox is the tedious
nature of procedures for bringing purified solvents (distilled into
special containers) in through the antechamber, and the shortage of
interior space in which to store them. These difficulties particularly discourage purification methods requiring large volumes of
solvent, such as preparative scale liquid chromatography.

Numerous solvent intensive procedures are facilitated by the
installation of direct liquid transfer feedthroughs. The assembly
illustrated in Figure 5 may be easily installed into either aluminum
or Lexan glovebox panels using only an electric hand drill.(9) The
primary components consist of a Swagelock(3) 1/8" OD tube to bulkhead male connector and a Whitey(10) 3-way Ball valve, and all
liquid connections employ 1/8" OD Teflon (5) or stainless steel
tubing.

A primary liquid feedthrough application is the anaerobic
transfer of purified solvent into the glovebox, either for temporary
storage or on an as needed basis. Solvents purified by distillation
(from drying and deoxygenation agents) may be transferred either
directly from the collection reservoir (if the still is located
somewhere in the vicinity of the glovebox) or from an external
storage bottle. Whatever the immediate external source, it should
be equipped with a 3-way valve, allowing selection between an inert
gas purge stream and the solvent, as illustrated in the upper
portion of Figure 6.

A typical solvent transfer operation proceeds as follows. The
external reservoir valve is turned to permit the flow of the inert
gas, which is vented externally at the feedthrough valve, thus
purging the transfer tube. With the inlet tube inserted into the
final solvent destination, both valves are turned to allow the
solvent flow into the glovebox. When the transfer is complete, the
external reservoir valve is switched back to allow the inert gas to
flush through the remaining contents of the tube, after which both
valves are turned off. The pressure differential required to effect
convenient transfer rates will vary, but in general the external
reservoir should be constructed to safely tolerate at least five
psig positive pressure.

A basic system for performing internal column chromatography
utilizing two direct liquid feedthroughs is illustrated in Figure 6.
While solvents transferred into the glovebox may be applied directly
to the column, it may often be more convenient (particularly when
using solvent mixtures) to employ an internal solvent reservoir.
Solvent delivery may then be accomplished either by applying inert
gas pressure over the reservoir or (for steady flow rates) by
employing a metering pump like the one used for recirculating
coolant.

Columns used for internal chromatography operations are most
conveniently dry-packed outside the glovebox. Both ends should be
fitted with a 3-way chromatography valves(11) to facilitate sample
injection and effluent distribution operations. After purging the

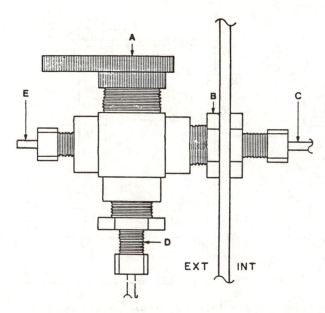

Figure 5   Direct Solution Transfer Feedthrough.   A: Whitey(10) 3-
way ball valve with 1/8 FPT port connections,
B: Swagelock(3) tube to 1/8 MPT bulkhead adapter,
C: 1/8" OD Teflon or stainless steel tubing,
D: Swagelock(3) tube to 1/8 MPT adapter, E: Cajon(8)
1/8 MPT to nipple adapter.

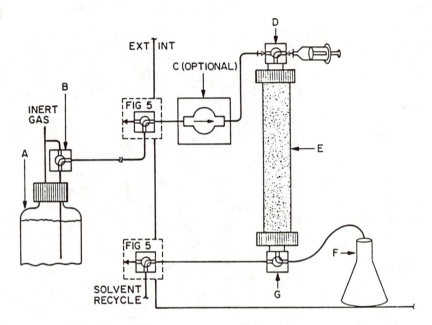

Figure 6   Apparatus For Internal Liquid Chromatography. A: still
           collection reservoir or storage bottle, B: 3-way valve-
           inert gas/solvent, C: adjustable pump, D: 3-way valve-
           sample injection/solvent, E: chromatography column,
           F: sample collection flask, G: 3-way valve-external
           recycle/sample collection.

assembly with a rapid flow of dry inert gas for at least one hour,
columns are transferred (with valves closed) to the glovebox ante-
chamber. Immediately before closing the antechamber door and begin-
ning evacuation, both valves are switched open. Due to the slow
outgassing behavior exhibited by most stationary phase materials,
the attainment of antechamber pressures of less than 5 millitorr may
take more than a day. This may often be unnecessary and good
results have been obtained by performing two one hour and one over-
night evacuation and refill cycles.

Reduced decomposition of highly air sensitive samples has been
noted if the dry column is preconditioned by flushing with three
column volumes of a polar, hydrophylic solvent such as tetrahydrofuran, followed by ten volumes of the selected eluent. The large
volumes of solvent used for this process, as well as those fractions
determined not to contain air sensitive products may be transferred
directly out of the glovebox using a second liquid transfer feedthrough, thus significantly decreasing internal space and container
requirements. In considering the use of solvent intensive procedures, it is important to emphasize the need to minimize the
accumulation of vapors in the internal atmosphere. Many common
solvents not only deactivate purification catalysts but cause crac-
ing of plastic glovebox panels, thereby greatly reducing their
resistance to breakage under stress. When open flasks are unavoid-
able (as in chromatography applications) containers may be placed
inside a large tabletop style polyethylene Buchner funnel.(12) The
funnel is connected to a regulated vacuum feedthrough which is
adjusted to provide a slow flow of gas and solvent vapors out of the
glovebox.

Applications and Discussion

The systems described in the preceding sections allow essentially
all operations normally employed in chemical synthesis to be performed entirely inside a glovebox. In many cases, air sensitive
reactions may be performed far more efficiently using less elaborate
glassware and with more rigorous exclusion of oxygen and water than
would be possible externally.

The installation of a recirculating heat exchange system ex-
pands the glovebox applications to include all reactions requiring
elevated temperatures and localized heat removal. The system described allows the convenient use of standard, normally water cooled
glassware (such as reflux condensers, sublimators, and distillation
apparatus) inside the glovebox. The ability to perform these operations internally becomes particularly useful in procedures involving
stepwise additions of air sensitive reagents (particularly of
solids) or the periodic removal of samples to monitor the progress
of a reaction. Rapid qualitative analytical techniques such as thin
layer chromatography are easily applied to internal reactions, but
almost impossible for air sensitive syntheses conducted externally.
In addition, the preparation of reaction aliquots for spectroscopic
analysis is greatly accelerated by eliminating the delay associated
with bringing the sample in through the antechamber.

Routine applications of the refrigerated immersion probe bath
include low temperature synthesis, collection flask cooling (for low
pressure distillations), and numerous manipulations involving

thermally unstable materials. Particular advantages exist for studies of low temperature photoreactions, in which the absence of condensing moisture allows the use of less complex illumination apparatus and facilitates observation and sampling. Finally, an additional benefit is the passive removal of solvent vapors from the internal atmosphere. Stray solvents simply condense on the probe and are collected in the bath, which is easily changed when necessary.

Direct liquid transfer feedthroughs greatly accelerate many internal operations by providing a quick alternative to transfers through the antechamber. In addition to liquid chromatography applications, such feedthroughs may be used for the direct addition of an external reagent to an internal reaction, a particularly valuable capability for working with thermally unstable materials which would decompose on the way through the antechamber. Conversely, they allow the quick transfer of internal samples into external cells for spectroscopic analysis.

In summary, the installation of these systems converts a standard glovebox system in to a versatile tool for chemical synthesis.

References and Notes

1. Shriver, D. S.; Drezdzon, M. S., "The Manipulation of Air Sensitive Compounds", 2nd Ed., John Wiley & Sons, New York, NY, 1986. b) Barton, C. J., "Glove Box Techniques", in Techniques of Inorganic Chemistry, Vol. III, Interscience, New York, NY, 1963. c) Walton, G. N., ed. "Glove Boxes and Shielded Cells", Butterworths, London, 1958.

2. Glovebox and accessories described in product information available from Vacuum/Atmospheres Co., 4652 West Rosecrans Ave., Hawthorne, CA 90250.

3. Bored through Swagelock fittings for use with feedthroughs must be specified using a "BT" after the appropriate catalog number. Available from Crawford Fitting Co., Solon, Ohio 44139 and through various distributors.

4. Fluid Metering, Inc., Box 179-T, 29 Orchard St., Oyster Bay, NY 11711.

5. Teflon and Viton are registered trademarks of the E. I. DuPont de Nemours & Co.

6. Stainless steel Swagelock (3) quick disconnects with Viton O-rings exhibited fair compatibility with hydrocarbons, but occasionally leak below 0°C.

7. Neslab Instruments, Inc., P.O. Box 1178, Portsmouth, NH 03801.

8. Manufactured by the Cajon Company, 9760 Shepard Road, Macedonia, Ohio 44506. An Ultratorr fitting is employed on the inside because it uses a compressed O-ring seal rather than a ferrule lock, allowing the entire probe assembly to be easily removed. For this application the bored-through modification must be specified.

9. The assembly illustrated mounts in a 21/64" hole and is best located where readily accessible both from the inside and outside (one hand in, one hand out), generally on a side panel. The bulkhead connector should be either permanently mounted with epoxy or sealed on the exterior side with an O-ring.

10. Whitey Co., 318 Bishop Rd., Highland Heights, Ohio 44143.

11. A suitable lightweight valve allowing solutions to contact only fluorocarbon polymers is available from the Hamilton Co., P.O. Box 10030, Reno, Nevada 89510.

12. Described in the equipment section of the Aldrich Catalog, Aldrich Chemical Co., 940 West Saint Paul Ave., Milwaukee, Wisconsin 53233.

RECEIVED September 24, 1987

Facile and Inexpensive Repair of Drybox Gloves

R. R. Schrock and A. H. Liu

Department of Chemistry, 6–331, Massachusetts Institute of Technology, Cambridge, MA 02139

Replacement of the hand portion of a drybox long-sleeved glove is a simple and cost efficient method for repairing the most common damages to these gloves. The procedure for this process and the procurement of the necessary parts are described.

The damage most frequently incurred to a drybox glove is the puncturing or tearing of the hand portion of the relatively expensive long-sleeved glove. Consequently, a glove with its arm portion in good condition is often replaced because of damage to the hand portion. Therefore, a method of replacement of only the hand portion is desirable.

Materials

The materials needed to convert normal drybox gloves into "two-component" gloves are ring and spring assemblies and medium weight 11" butyl gloves. The gloves are available in sizes 7 to 11 but the ring and spring assemblies are available in only one size. In our experience, the rings have been sufficiently large to accommodate all hand sizes; the only case where they were not large enough involved a broken wrist in a cast. These materials are probably available from most rubber glove suppliers but are definitely available from University Rubber Co., Inc.; 42 Brookford St., Cambridge, MA 02140; (617)864-9733.

Procedure

Conversion. Cut off the hand portion of the long-sleeved glove at its narrowest point, approximately three inches above the thumb. Tape the ring from a ring and spring

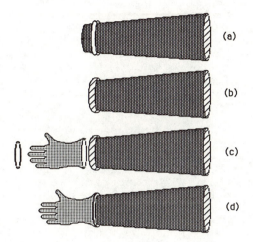

Figure 1. Conversion of a normal long-sleeved glove to a two-component glove.

assembly with one or two windings of electrical tape and
slip it onto the now tube-like sleeve as shown in Figure
1a. Roll the sleeve back onto the ring (the sleeve
should fit snuggly) and, like before, tape the sleeve
covered ring with one or two windings of electrical tape
(Figure 1b). Next, slip a 11" butyl glove onto the
wrapped ring and once again protect the glove with one or
two windings of electrical tape (Figure 1c). Secure the
glove with the spring as shown in Figure 1d. Tape the
spring, now assembled with the ring, sleeve, and glove,
with three or four windings of electrical tape. The
conversion is now complete. Note that all the glove and
sleeve surfaces are protected from the metal parts by
electrical tape. To convert a glove that is on an
operating drybox, read the replacements section below for
isolation and purging instructions.

Replacement. To replace a damaged glove, first mark the
position of the thumb on the sleeve with a marker.
Isolate the damaged glove by twisting the sleeve and
creating a constriction in the middle of the sleeve.
Secure the constriction with electrical tape. Remove the
spring, and replace the glove with the thumb lined up
with the mark on the sleeve. Before the new glove is
secured with the spring, however, the air trapped in the
lower sleeve and glove should be flushed out. Fill the
glove with nitrogen by slowly bleeding it in from the
drybox through the constriction (the tape securing the
constriction may need to be loosened slightly) and, while
separating the glove from the sleeve, quickly force out
the nitrogen and air. This procedure should be repeated
several times and the spring replaced as described above.

Acknowledgments

We would like to thank all the former members of the
Schrock group who contributed to the development of this
technique.

RECEIVED July 31, 1987

Chapter 6: Application 2

Continuous Inert Gas Purge Glovebox for Organometallic Synthesis

Todd B. Marder

Guelph–Waterloo Centre for Graduate Work in Chemistry, Department of Chemistry, Waterloo Campus, University of Waterloo, Waterloo, Ontario N2L 3G1, Canada

There is no question that inert atmosphere glove-boxes represent a very useful tool in organometallic chemistry, complementary in nature to Schlenk and high-vacuum techniques. One limitation which has been encountered by many researchers is that the use of certain classes of organic solvents (eg. halogenated ones) or reactive ligands containing phosphorus or sulfur react with the deoxygenation catalyst in the atmospheric recirculation system. This problem has led many groups to limit the use of their glove-boxes to chemistry not involving these reagents or solvents, or, in certain cases, to avoid performing anything other than solid storage and transfers in the box.

A second factor which I have observed is the lengthy pump/refill cycles often employed for ante-chamber atmosphere purification requiring the chemist to wait 20-30 minutes before bringing new reagents or equipment into or out of the box. These two limitations are sometimes perceived as inherent limitations to the general technique of glove-box manipulations and provide an inhibition to the use of such equipment.

We describe below some extremely simple solutions we have encountered while working at the DuPont Central Research and Development Department and which we currently employ in our research laboratory.

The most important single issue is using the "right tool for the job" and this requires making an informed decision about the nature and reactivity of the compounds to be manipulated. In particular, one has to ascertain the sensitivity of one's compounds to oxygen and to water. The term "air-sensitive" is a catch-all phrase which does not distinguish between oxygen and water, although the two are often treated as a single problem. Our experience with highly-reactive, low-valent phosphine complexes of Rh and Ir highlights this distinction. For example, whereas both $[(PMe_3)_4Rh]Cl$ and $[(PMe_3)_4Ir]Cl$ are quite oxygen sensitive, only the latter reacts irreversibly with water (1), the former is stable (and soluble) in deoxygenated water and we have carried out numerous stoichiometric and catalytic reactions using this property. Clearly there are also compounds which react rapidly with water but not necessarily with dry O_2. Finally, one needs to discern how sensitive is "sensitive"?

0097–6156/87/0357–0153$06.00/0

We find that many "air-sensitive" compounds can be satisfactorily stored and manipulated in a glove-box which has <u>no</u> recirculation unit whatsoever, but instead uses a continuous <u>slow</u> purge of N_2 gas. This is supplied at ca. 50 psi as the boil-off from a large liquid N_2 storage facility. The gas is currently used for both the purge and the photo-helic controlled box-pressure system via independent local pressure regulators. The purge gas exits from the glove-box via a 3/8" copper tube (attached to one of the fittings supplied as standard equipment with the glove-box) into a large-bore Hg bubbler located in a fume hood adjacent to the unit. Our flow rate is limited by the 3/8" bore size of the exit tubing and the pressure is maintained at ca 2" of H_2O (see the Photohelic gauge) by only a few millimeters of Hg below the bubbling tube. A large volume bubbler (ie. wide) is used to eliminate the possibility of back flow of air into the glove-box via the bubbler in the event that the pressure inside drops slightly below atmospheric during evacuation or refilling of the ante-chamber. We find that a flow rate equivalent to a little over one-full size N_2 cylinder per week is adequate for our needs (2). The quality of the atmosphere is checked routinely using toluene solutions of Et_2Zn and Me_3Al as O_2 and H_2O indicators respectively. Some "smoking" of the Me_3Al is always evident but there is almost no detectable "smoking" of the Et_2Zn solution. By comparison, neither reagent should smoke in a glove-box equipped with a recirculation train containing O_2 and H_2O scrubbers. However, the relatively low levels of O_2 and H_2O in our atmosphere (limited by the quality of our liquid N_2 boil-off) (3) are acceptable for <u>our</u> needs. One could improve the gas quality with various scrubber units in series with the purge stream but <u>we</u> have not found this necessary. This setup allows us to keep ca 1ℓ capped bottles of solvents as well as vials of liquid phosphines (eg. PMe_3) in our box and to manipulate these with disposable pipets. Small scale reactions (ca. 5g) can often be conducted in standard glass liquid-scintillation vials (ca 25 ml) which can be disposed of after use, used for sample storage, recrystallization, etc. We employ a low-profile digital balance (1 mg sensitivity) (4), small magnetic stirrers and several ring stands as standard equipment. In addition, we have built shelving units for the back wall and end panel to our own specifications to make the most use of the limited space available (5). We use stackable plastic bins of various sizes (which fit our shelves by design) and store hundreds of sample vials and small reagent bottles, spatulas of different sizes, syringes, disposable pipets and rubber bulbs, empty vials, small round bottom and filter flasks, fritted and regular funnels, 5 and 10 mm nmr tubes, deuterated solvents, plastic, rubber and ground glass stoppers, magnetic stir bars, etc. An external vacuum-pump with double solvent trap unit (one ca 1ℓ trap for solvent and a second standard trap for added pump protection) is connected to the back of the double-length glove-box via two rear ports with valves located inside the box. Attached to the valves are thick-wall rubber hoses and via two metal T-tubes, each line is split into three lines, with an in-line stopcock on each. These are used for filtrations, for solvent stripping (we also have a small stainless rotary evaporator) (6) and quick sample drying. For crystallizations and storage of thermally sensitive materials, we have a built-in freezer unit operating (7) at ca -35°C.

Rapid entry or removal of materials and supplies from the box are facilitated by the use of a large (845 ℓ/min) vacuum pump distinct from the smaller pump discussed previously. The rapid pumping speed allows three ante-chamber pump-refill cycles to be completed in just over three minutes total time.

The facility described above allows for <u>rapid</u> and <u>convenient</u> performance of <u>routine</u> manipulations of moderately O_2 or H_2O sensitive materials. It is complimentary to Schlenk and high-vacuum lines and <u>all</u> are utilized in our lab depending on the specific requirements of the particular task.

A final note concerning safety seems appropriate. It is critical that all vapors in the glove-box be properly vented in a fume hood. Thus, as described above, our purge gas exits into an Hg bubbler <u>in the fume hood</u>, and we have cut away part of the side of a fume hood in order to insert the glove-box ante-chamber face well inside the hood. Remember, when removing items from the glove-box, the ante-chamber will contain the same atmosphere as inside the box. You cannot smell the toxic fumes of even PMe_3 when properly contained inside the box but such vapors will be inside the ante-chamber after the inner door has been opened! The ante-chamber is equipped with a sliding tray to facilitate loading and unloading. Simply slide the tray out rather than having to put your face near the outer door even in the fume hood.

REFERENCES

1. D. Milstein, J.C. Calabrese and I.D. Williams, <u>J. Am. Chem. Soc.</u>, **1986,** 108, 6387; T.B. Marder, D. Zargarian, J.C. Calabrese, T. Herskovitz and D. Milstein, submitted; T.B. Marder, D.M.T. Chan, W.C. Fultz and D. Milstein, 12th International Conference on Organometallic Chemistry (ICOMC), Vienna, Austria, September 1985, Abstract # 163; D. Milstein, T.B. Marder and W.C. Fultz, ibid., Abstract # 369.
2. Boil-off from a large ℓN$_2$ facilty provides a much less expensive source of N$_2$ gas than cylinders.
3. Gas quality varies with supplier and facility.
4. Static, a major nuisance in a dry environment, can be controlled with the use of a Zerostat gun available from Aldrich. It is important to discharge sample vials during taring and weighing operations.
5. The standard shelf units supplied by Vacuum Atmospheres Co. do not make the best use of available space. We suggest building one's own shelves to suit specific needs.
6. Rinco rotating evaporator, manufactured by Valley Electromagnetics Corp., Spring Valley, Illinois.
7. Model DC-882 Dri-Cold freezer available from Vacuum Atmosphere Co., Hawthorne, California, attached to our Vacuum Atmospheres HE-553 Dri-Lab glove-box. We and others have experienced unreliability in the compressors currently being supplied with the DC-882 freezer. The one-year standard warranty is helpful. When operating, the freezer is an excellent and most useful piece of equipment. Vacuum Atmosphere Co. is aware of the problems and is redesigning the refrigeration system. Other manufacturers also supply glove-boxes with freezers.

RECEIVED July 31, 1987

Chapter 6: Application 3

An Inexpensive System That Enables the Cooling of Reactions Inside a Glovebox

Jeffrey Schwartz and Kevin Cannon

Department of Chemistry, Princeton University, Princeton, NJ 08544

Work with air- and moisture-sensitive organometallic compounds is often facilitated by the use of a glove box. However, manipulations which require reaction refluxing are often not readily amenable to use in a glove box because it is not easy to cool the appropriate apparatus inside the box given obvious hazards associated with the use of water as a cooling medium. Also, running reactions in the cold inside a glove box can often entail the use of expensive cooling equipment. We describe here an inexpensive cooling system compatible with requirements of the glove box. This system is easy to build and easy to operate.

Most of the components of the cooling system described are located on the outside of the glove box to enable easy access to such components and to minimize clutter inside the glove box (see figure 1). In this system, methylcyclohexane is used as the coolant fluid. This material can be purified before its introduction into the cooling system such that no contamination of the glove box by water or oxygen would occur in the event of a cooling line rupture. The purified methylcyclohexane is added to the reservoir (3) through the filling port (5); nitrogen is then bubbled through the reservoir to place an inert gas blanket above the medium. The "swagelok" port is then capped. The reservoir in the system should be opened only if additional coolant is needed in the event of an inadvertent loss. The reservoir is constructed from a standard ether can equipped with welded "swagelok" fittings (3/8"). All tubing outside the glove box is 3/8" OD aluminum, and all connections are made via "swagelok" fittings. Inside the box, tygon tubing is used.

To activate the cooling system, connect the apparatus (9) (either a reflux condensor or a cooling loop) to the tygon tubing taking care to note the inlet and outlet tubes. A cooling bath is then placed around the aluminum cooling coil (1) at the appropriate temperature desired. The flow control valve (6) and both shut-off valves (7,8) should be in a closed position at this time. The flow control valve and the by-pass valve are Whitey regulating valves

0097–6156/87/0357–0156$06.00/0
© 1987 American Chemical Society

(#B-1RS6; $17.00) and the shut-off valves are Whitey ball valves (#SS-44S6; $79.00). After opening the by-pass valve (4), one can then turn on the circulation pump (2). The pump used in our system is Little Giant, Model 2E-NDVR, which was obtained through Industrial Electric (#502303; $79.40). At this point, the methylcyclohexane will be circulating from the reservoir to the cooling coil, and back to the reservoir by way of the pump.

To circulate coolant through the apparatus, open both shut-off valves and slowly open the flow control valve, adjusting it to the proper flow. If completely opening the control valve gives insufficient circulation, slowly closing the by-pass valve will further increase the flow. It is important to remember that the flow rate will vary according to the temperature of the coolant.

To deactivate the cooling system, fully open the by-pass valve and turn off the pump. With the flow control valve completely opened, residual coolant in the apparatus will be siphoned back through the inlet port down into the reservoir. Both shut-off valves are then closed. After the apparatus is disconnected from the tygon tubing, the two ends of the tubing can be mated and left in place.

NOTES

A reviewer has suggested the use of "Special Fuel Grade" or "Viton" tubing in place of Tygon tubing, and also locating the pump inside the glove box, with only the cooling loop located externally.

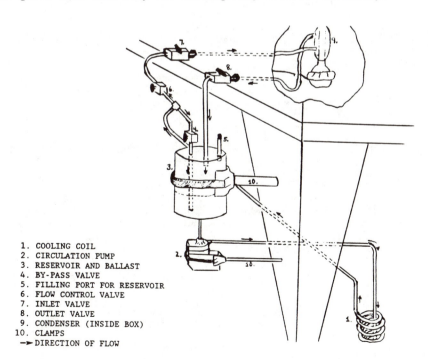

1. COOLING COIL
2. CIRCULATION PUMP
3. RESERVOIR AND BALLAST
4. BY-PASS VALVE
5. FILLING PORT FOR RESERVOIR
6. FLOW CONTROL VALVE
7. INLET VALVE
8. OUTLET VALVE
9. CONDENSER (INSIDE BOX)
10. CLAMPS
→ DIRECTION OF FLOW

RECEIVED September 24, 1987

Chapter 7

Using Metal Atoms and Molecular High-Temperature Species in New Materials Synthesis

Apparatus and Techniques

Mark P. Andrews

AT&T Bell Laboratories, Murray Hill, NJ 07974

This article describes recent advances in the area of preparative scale vacuum evaporation technology which have greatly extended the range of application of free transition metal atoms and molecular high temperature species for chemical synthesis. These advances include (1) the development of liquid nitrogen-cooled electron beam furnaces for chemical compound synthesis in cooled, liquid organic solvents; (2) the use of large cooling capacity helium gas cryostats for synthesis under quench condensed conditions at temperatures below 20 K; (3) techniques and apparatus for using photoexcited metal atoms and molecular species for preparative scale synthesis of organometallic compounds; (4) and procedures for examining the microscopic events leading to chemical reactivity and stability which are important in designing large-scale reactions. Fundamentals of design, principles of operation, reactant and product manipulation, quantification, and applications for new materials synthesis are elaborated.

Technology evolves dynamically through the development of new or improved materials, which in turn permit the creation of new technologies. Successful technology is usually crucially linked to leadership in materials science where advances often arise out of the close collaboration of physicists, chemists and engineers. The chemistry of substances available for use as materials continues to develop at the confluence of assessed physical properties, chemical

0097–6156/87/0357–0158$09.00/0

reactivity, end use (in materials science and device fabrication), and advances in synthetic techniques. Among these synthetic techniques are those which utilize atomic or molecular vapors generated in a vacuum and condensed at a cold surface where products may be formed and isolated. This is the principle of some of the thin film technologies - it is also the elementary principle behind the present day use of free atoms of the transition metals, lanthanoids, and actinoids, and molecular high temperature species (HTS) in the preparation of organic, organometallic, co-ordination and metal-cluster compounds and catalyst materials.

The introduction of the carbon-vapor reactor(1) initiated the use of high temperature species for synthesis. Extensions to the metallic elements followed.(2-4) The rather extensive literature (1-12) that has issued from but a small number of laboratories, recognizes the kinetic and thermodynamic advantages of this technique in providing more direct or higher yield routes to existing compounds and making available new compounds and materials. The use of atomic and molecular high temperature species for chemical compound synthesis is well documented,(2-8,13,14) and there is a practical exposition of techniques for dealing with low volatility compounds and recovering air sensitive materials following this chapter. Applications of vapor synthesis (VS) to materials are really just emerging. A brief overview of the subject has been given.(12) Thus, besides its utility in making organic and organometallic compounds, vapor synthesis has seen expression in polymer science, catalysis, colloid synthesis, and electro-active materials. Some examples are provided at the conclusion of this chapter.

Recent design sophistication in the area of preparative scale vacuum evaporation technology has introduced additional synthetic flexibility and greatly extended the range of application of vapor synthesis. These advances reflect changing requirements in the field of vapor synthesis, paired with a consciousness of the potential for contributions to new materials synthesis.(12) This chapter sketches these advances in terms of design fundamentals, principles of operation, reactant and product manipulation, quantification and synthetic strategies.

Experimental Approaches

VS techniques can be distinguished on the basis of whether the reaction vessel is stationary or rotatable.(2-6) The procedures are further distinguished according to whether all vapors are condensed at low temperatures to give a solid, fusible film (static VS), or whether metal atoms or HTS are deposited directly into dynamically mixed neat liquids, solutions or suspensions of solids (liquid phase VS). Given the same starting materials, the products obtained may reflect the choice of technique or procedure. Except in the intended preparation of metal aggregates, the facile self-association reactions of the condensed vapors must be suppressed to favor atom co-reactant interactions; hence, high pumping speeds are required with sensitive strategies for mixing reagents, as are efficient furnace and refrigeration designs.(3-6)

Cocondensation Methods Using Static Reactors(4-7,10), Background.

The glass or stainless-steel reactor is the cheapest to fabricate
and easiest to use.(4-7) A bell jar version of such a reactor is
depicted in Figure 1, whereas a version which is submersible in a
cryogen such as liquid nitrogen is represented in the Figures in the
short article following this chapter. Vessels of 1-50-L capacity
can be manufactured to accommodate furnaces of various designs, ie,
resistance, electron-beam, laser, induction, or electric arc heating, and cathode or magnetron sputtering. Procedural details per-
taining to the synthesis of organometallic compounds in static glass
reactors are provided in references(6,10 and 14). Resistance heating is suitable for evaporating Cr, Mn, Fe, Co, Ni, Pd, Cu, Ag, Au,
Zn, Ge, Sn, Pb, elements of groups 5 and 6, and a multitude of
molecular species.(5-10) To reduce the sometimes deleterious
effects of thermal radiation from crucibles, we have found it
advantageous to enclose them in tantalum radiation shields.
Alternatively, refractory wools can be wrapped about the crucibles,
however, in some cases pyrolysis on the large surface area can
prevent reaction or cause free radical polymerizations of
alkenes.(15) In larger than milligram quantities, Ti, V and the
more refractory materials are best evaporated by electron
beam.(13,14) In any case, for metals it is usual to degas the
element in a preliminary step by melting or subliming. In a
separate step, the metal surface is degassed just below its
vaporization point until a stable base pressure is reached without
cooling the vacuum vessel. Heating is reduced and the cryogen is
installed. Normally, the reaction chamber is isolable from the
vacuum station, which is capable of sustaining a pre-reaction vacuum
of $<10^{-3}$ torr. The (degassed) co-reactant is introduced as a neat
liquid, diluted in a solvent that is inert to reaction, or coinci-
dently with a separately cannulated solvent. (These manipulations
are most conveniently carried out by means of an independent source
of vacuum and inert gas from a nearby Schlenk manifold.) The metal
is gradually brought to the desired vaporization rate and continuous
slow cocondensation is maintained for 1-4 h. Upon termination, the
pumping system may or may not be isolated, depending on the need to
remove volatile, unreacted material or to introduce secondary reagents intended for reaction upon warm-up. Usually, the solid
matrix is melted slowly within the reaction vessel under an inert
gas atmosphere. Resulting fluids are immediately cannulated and
filtered to remove active metal particles which might decompose the
products.

Reverse-Polarity Electron Beam Sources.(13,14)

There have been some important modifications to electron beam furnaces which have made them more suitable for VS.(13,14) For instance, the range of applicability of electron beam vaporizations
has been extended to liquid phase systems through the development of
a liquid nitrogen cooled hearth.(13) This feature will be discussed
later (see Rotating Reactors: Liquid Phase Methods). For use in
static (and rotating) reactors, electron beam sources have been
adapted in general to reduce damage to molecules by parasitic ions,
and secondary (scattered) electrons, and to overcome the difficulty

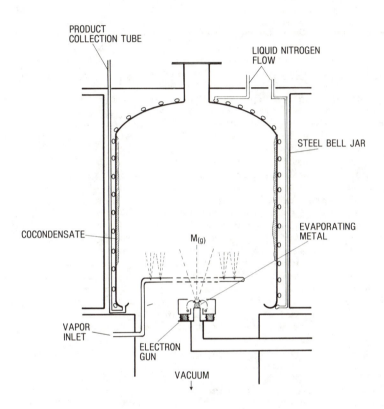

Figure 1: Steel bell jar static VS reactor for electrostatic, reverse polarity electron beam evaporation and 77K cocondensation.

introduced due to plsama discharges when one requires uninterrupted
evaporations in potentially high vapor-pressure environments. These
requirements have been met by operating a work-accelerated, electrostatically focused electron gun in a "reverse polarity" mode, where
the water-cooled hearth and its metal sample are held at a high
positive potential. This arrangement suppresses electron scattering
and reduces ion emissions to microamp levels(13,14) that are tolerable during synthesis. For those unfamiliar with the principle of
operation of such an evaporation source, a short description
follows.

Principle of Operation.

Units incorporating multiple vaporization sources provide obvious
advantages in multi-element preparations.(13,14,16,17) A close-up
of a dual reverse-polarity, electrostatically focused electron gun
source is illustrated in Figure 2. The furnace is similar to that
manufactured by Torrovap Industries, Inc.(16) The interchangeable
hearths are set against o-rings seals in the base of the water cooling stages. This arrangement facilitates modifications of the
hearth geometry, adding flexibility in adapting the hearths to
unique experimental requirements. The DC circuit required for oper-
ation(18) is driven by two three-phase (208/220V) high voltage power
supplies. These are variable from 0-10 kV at 0.2 A. The tri-phase
fullwave rectification produces a DC output voltage with about 14%
peak-to-peak ripple, eliminating the need for filter capacitors
which can retain charge. This arrangement is less destructive to
the electron gun and safer for the operator. A triac is incorporated to control the input voltage and DC output for each gun. If the
current exceeds 0.2 A the input power is instantaneously cut by a
de-energizing self-holding latch in the HV circuit. This event
activates an HV circuit breaker to disable the guns. The triac
circuit is operated by a 10 V power supply, floated to permit true
DC triggering and to shield the triac from false triggers due to
transients in the line neutral.
   An AC power supply (30 A max.) provides the filament current to
the respective electron guns. The thoriated tungsten filament has a
work function of about 2.6 eV. Those of W and Th are 4.52 and 3.35
eV, respectively. (Filaments for electron guns can be cheaply fabricated. Thoriated tungsten (1% ThO_2) filament wire, 0.015 in.
dia., can be obtained from General Electric.) Electrons thermally
emitted from the filament find themselves moving along the lines of
force curving towards the hearth in the electric field established
by the high positive potential at the copper anode. The filament
and metal target must be carefully positioned to ensure that the
electrons moving in the accelerating potential will converge at a
point to surrender their kinetic energy ($E=eU_B$, where U_B is the
accelerating voltage) to the metal to give the highest power density. The design of the electrostatic lens recalls one developed
originally Unvala.(19) In his apparatus both the focusing shield
and the filament are held at a high negative potential. Electrons
are directed to a grounded target. In the design in Figure 2, components other than the cathode and anode in the hearth area are at
or near ground potential. Like the Unvala design, there is no line

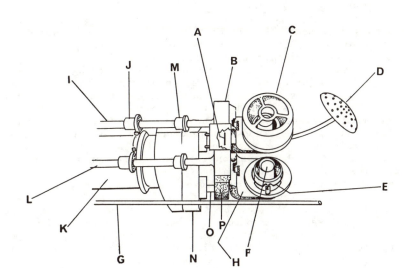

Figure 2.   Close-up of twin electrostatic, reverse polarity
electron beam furnace.   A: Quartz crystal mass monitor; B:
Liquid nitrogen or water-cooled copper stage; C: Stainless steel
gauze electrostatic focusing can; D: Heated ligand inlet shower
head;   E: Thorium doped tungsten emitter filament; F: Water-
cooled copper hearth; G: Cannula; H: Adjustable stage for supporting focusing can and emitter filament; I: Liquid nitrogen or
water inlet/outlet tube; J: Teflon support ring; K: Stainless
steel shaft housing high/low tension wires and water cooling; L:
Liquid nitrogen/water inlet/outlet tube; M: Teflon collar; N:
Glass-filled Nylon feedthrough; O: Water connection for copper
hearth;   P: Liquid nitrogen/water-cooled copper stage.

of sight path for electrons to travel from the filament to the metal
target. This arrangement therefore reduces the chances of improper
focusing and the deleterious effects of contamination and sputtering
of the filament by adventitious ions.

The hot emitter filament gives an emission current density
I_{em} determined by its temperature and described by the
Richardson-Dushman equation, $I_{em}=AT^2\exp(-B/T)$, where A is the
Richarson constant, and $B=eU/k$ reflects the work function of the
metal (k is Boltzmann's constant). Both A and B depend strongly on
the choice of metal, the most common ones being W, LaB_6 and Ta. The
saturation current density of the filament is also affected by the
field strength in front of it. Should the field strength be too low
to extract the majority of electrons, an electron cloud forms about
the filament and the resulting space charge depresses the cathode
emissions. Use of small area cathodes generally enhances the space
charge limited current density. There is a transition range between
the space charge and temperature dependent saturation emission
current densities. The cathode is usually operated within this
transition to obtain an emission current density at a cathode temperature that is as low as possible.

Since the gun behaves as a diode operating in the saturated
regime, the beam current is controlled solely by the electron emission from the filament (ie, by the filament temperature). Beam
focus is rather sensitive to the filament loop diameter and slightly
sensitive to the vertical position of the filament. Optimum positioning can be established by trial and error. Additional focusing
can be achieved by applying a small DC bias to the electrostatic
focusing shields. This bias helps to concentrate the beam on the
diminishing bead of metal throughout the vaporization. Ideally, all
field influencing components in the construction of the electron gun
should observe exact rotational symmetry in order to avoid axial
astigmatism.

Static Reactors with Electron Beam Capabilities.

Figure 1 shows a schematic bell jar reactor for electron-beam evaporation for chemical synthesis. Commercial static reactors(16,17)
similar to that depicted in Figure 1, now offer single and dual
electron guns (power rated at 2-10 kW), hybrid resistance/electron-
beam furnaces which can be accommodated in glass or stainless-steel
bell jar reactors, and modular designs to permit various combinations of reaction vessel sizes, pumping stations and furnaces. For
example, a vapor synthesis plant (Type 500) offered by G. V. Planar
Ltd.(17) (cf. Figure 1) features a 12-18 in. dia. insulated
stainless-steel bell jar with integral liquid nitrogen cooling. A
gutter at the base of the vessel enables products to be collected
and removed. An access port is provided at the top of the vessel.
The vapor sources are single or dual, reverse polarity units with
interchangeable water cooled hearths. The sources together with
water and electrical connections are mounted in a service ring. The
stabilized high voltage power supply is continuously variable up to
7 kV/0.5A+0.5A, with the high tension common to both sources or
independent. Filament supplies for independent operation of the
dual furnace are each rated at 24V/70A. The circuitry is surge and

overload protected and the vacuum system is interlocked prevent
inadvertent operation. An electrically heated gas ring at the base
of the jar is used to distribute the ligand vapor. The pumping
station consists of a 160 mm dia. "Diffstak" silicone oil diffusion
pump(20) with an integral water-cooled baffle, backed by a rotary
vacuum pump. Vacuum connections are established by manually operated high vacuum valves and a butterfly valve for isolating a demountable liquid nitrogen trap from the reaction vessel.

Electron Beam-Vapor Synthesis Below 20K.

An important innovation in vapor synthesis has led to the introduction of a preparative scale static vapor reactor which allows reactions to be conducted at temperatures below 20 K.(21) Cooling is
achieved by means of a closed-cycle helium gas refrigerator (4 W at
10 K, 70 W at 77 K, Air Products Displex 204), housed in a
stainless-steel vacuum chamber (Figure 3). In this unit, the cryotip evolves into a copper bowl whose concave surface can be rotated
through 180° into position directly above a centered, 3.5 kW
reverse-polarity electron gun. Exploratory work has concentrated on
the direct synthesis of transition metal carbonyls. Cocondensation
reactions involving V, Cr, Mn, Fe and Ru typically involved
deposition of 10-100 mg of metal vapor with 10-100 g of CO for
periods of 1-6 h. Products are cannulated through the top of the
reaction chamber via a triple-walled temperature-controlled (77-300
K) transfer tube. The original study(21) concluded that a higher
cooling capacity refrigerator is required to raise the yield of
materials to the gram scale for a reasonable deposition period.
Subsequent improvements in the cryoshielding of the electron gun
furnace have dramatically lowered the operating temperature of the
cryotip to 12 K and 17 K for Pd/CO or Pt/CO cocondensations,
respectively.(21) Significantly, it is estimated that for similar
experimental configurations, the capital cost of the closed cycle
helium gas system is recovered after about 200 runs when compared to
the cost of liquid helium alone.
    The development of the sub-77 K apparatus raises two important
points: In the first place, the device makes accessible a range of
chemically and physically interesting molecules hitherto restricted
to microgram-scale matrix isolation synthesis.(5,6) These molecules
are potentially valuable as precursors for new materials synthesis.
It is conceivable that compounds might be synthesized which would
show stability above 77 K, but which might require sub-77 K conditions for their synthesis from "non-condensible" gases. For example, $Ni(N_2)_4$ which must be synthesized below 77 K, decomposes in
the range 80-100 K.(21) This observation raises the second point
which relates to metal atom self-aggregation processes in the condensed phase. Generally, only when the temperature of a matrix is
lowered below about one third its melting point (the Tamman temperature) can well-defined metal atom-ligand interactions dominate the
aggregation of metal atoms which give ill-defined colloidal metal
compositions, viz., $Ni + N_2 \rightarrow N_x(N_2)_{chemisorbed}$. On the other hand a
judicious choice of matrix material can allow one to conduct
reactions with non-condensibles at elevated temperatures. (22,23)

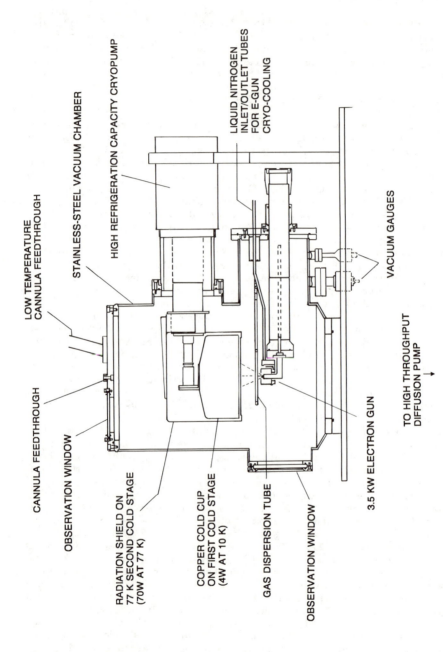

Figure 3. Schematic of a sub-20 K static reactor for preparative scale vapor synthesis by electron beam heating.

$$Cu + O_2 \xrightarrow[\text{adamantane}]{77K} Cu_3(O_2) + Cu_5(O_2) \qquad (1)$$

$$Co + C_2H_4 \xrightarrow[\text{methyclohexane}]{77K} Co(C_2H_4)_3 \qquad (2)$$

Although reaction 1 has been achieved only on a microgram scale, the fact that such exotic molecules can be isolated uner 77K cocondendation conditions makes adamantane a tantalizing high temperature matrix for low temperature synthesis.

Cocondensation Methods: Rotating Reactors(6,7,8,14)

A rotatable reactor for synthesis using low temperature cocondensation was introduced by Green and co-workers to provide better separation of the evaporate and substrate vapors in order to reduce pyrolysis of the latter at the furnace.(14) As shown in Figure 4, the electron gun (or some other furnace) is used to deposit a layer of atoms into a pre-deposited layer of substrate on the inner top surface of the rotating flask. In this manner, layer upon layer of solid accumulates. Isolation of the products follows procedures like those developed for the static devices.

Experience reveals that the rate of deposition of metal atoms and other high temperature species on a cold surface can have a marked effect on product yields. For applications requiring run to run reproducibility and a quantitative measure of evaporation rates, a mass monitor of the type used in thin film deposition is useful. An appropriately radiation shielded crystal can be located strategically to best sample the evaporate. The placement of such a sensor is given in Figure 2 for a commercially available dual electron gun rotary reactor.(16.17) The crystal is protected by a thin tantalum radiation shield set directly behind the hearths where it achieves temperature stability by being part of the liquid nitrogen (or water) cooled block (vide infra). The thin quartz wafer receives the flux of metal vapor through small holes drilled through the radiation shields. Choice of hole size reflects the desired mass resolution and the temperature coefficient of the crystal for a given metal. The crystal has the property that it can be made to undergo mechanical oscillations when supplied with electrical energy (piezoelectric effect).(24) The resonant oscillation frequency depends on the thickness of the crystal, or alternatively, on the mass of foreign material deposited per unit area on the crystal surface. It can be shown(25,26) that the rate of metal deposition is linearly related to the rate at which the oscillation frequency of the crystal decreases. Typically, absolute measurements of areal mass density are accurate to about ±2%. The limit of detection is about 20 ng-cm^{-2}.

Rotatable units offering various options for vapor sources, reaction vessesls, vacuum stations and ligand inlet configurations are offered commercially.(17,18,27) The Planar apparatus(17) utilizes a variable pitch, 5°-45°-horizontal 10 L glass flask which accepts single, twin and mixed vapor sources like those provided for its static models. The hybridized resistive/electron beam furnace

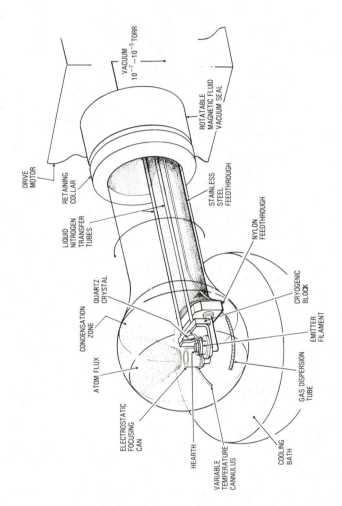

Figure 4. Schematic of a rotatable metal atom reactor for dual electron beam vaporization and cocondensation or liquid phase reaction.

offers a distinct advantage for simultaneous crucible-dependent and containerless evaporations.

Units manufactured by Torrovap Industries(16) are available in two sizes. The laboratory, bench-top model incorporates resistance or electron beam capabilities in 3-5 L flasks, quartz crystal mass monitoring, provisions for low temperature cannulation of temperature sensitive materials and safety interlocks. Like the smaller Kontes(27) apparatus, the device offers the advantages of ease of operation, quick turnaround associated with fast pumping of the reduced volume and surface area, and can be more easily scaled to work with smaller quantities of precious materials. The large model TVP-800 reactor unit is a safety interlocked modular structure whose electronics and vacuum station are designed to permit either static or rotary experiments with single or dual evaporation sources. The system is evacuated by a 2400 L-s^{-1} diffusion pump, backed by a direct drive mechanical pump. The vacuum head, located above the diffusion pump and gate valve, is a cylindrical chamber having a 6 in. side port for connection to a butterfly valve. A demountable liquid nitrogen trap extends into the top of the vacuum head. The main T-shaped vacuum chamber is connected to the butterfly valve on the side port of the vacuum head. One end of the T-flange holds the Plexiglass main flange which supports the electrodes for resistive heating. The Plexiglass flange is easily interchanged with a stainless-steel flange for mounting a single or dual electron gun. Opposite the main flange are the rotary seal drive and rotatable flange for the reaction flask. Like the Planar system, the rotary seal is made by means of a commercial ferrofluid. Such seals have proven to be reliable and robust after six years of use in our hands. The resistive evaporation source comprises three water-cooled copper electrodes arranged in a coaxial triangular configuration. The apical electrode serves as a common electrical ground. There is ample room for crucibles and radiation shielding in this set-up. Power to the electrodes is supplied by two 6 V/300 A and 16 V/60 A transformers. A quartz crystal mass monitor is also provided. The electron gun is similar to that described above (see Reverse-Polarity Electron Beam Sources, Principle of Operation).

Rotatable Cold Fingers(28-30).

Features of the rotating cryostat developed in 1961 for microsynthesis(28) are latent in the rotatable cold fingers introduced by two groups.(29,30) The apparatus described by Lagow and coworkers(29) has been used to synthesize $Cd(SiF_3)_2$ at 77 K by cocondensation of Cd atoms with trifluorosilyl and trifluoromethyl radicals generated by means of a low energy radio frequency source. The entire reactor can be rotated into a vertical position and the matrix can be collected in a refrigerated reservoir from which the products can be abstracted (Figure 5A).

Most metal atoms show a multitude of transitions into excited states, some of which are reactive towards substrates that are otherwise inert to the ground state atom. The use of excited state atoms and molecular species in preparative scale vapor synthesis offers tremendous potential for producing new chemistry. A rotating drum cryostat (Figure 5B) introduced by Billups and collaborators-

(30) includes a 450-W medium pressure Hg arc lamp for simultaneous photolysis of the condensate. The 60% yield of acetaldehyde resulting from irradiation of a chromium atom-methyl acetate matrix having a Cr:substrate ratio of roughly 1:700, indicates that reaction has occured involving Cr atoms shuttling back and forth between a photo-excited Cr-methyl acetate adduct and acetaldehyde.

Liquid Phase Methods: Static and Rotating Reactors (3-7,10,12)

Involtatile or temperature sensitive ligands, uncertainties in solid or fluid phase reactivity resulting from condensed state inhomogeneities, may make it desirable to conduct reactions in the liquid phase. Such reactions can be performed in static reactors by deflecting the vapors of some metals downward into the well-stirred solution.(10) However, a rotatable liquid phase reactor(2-4) can give greater flexibility (cf Figure 4). Such a device is operated according to the same principles described earlier for the cocondensation reactor except that vapors are directed upward into a thin film of fluid which is spun out of a turbulently mixed liquid, cooled to maintain a vapor pressure less than $<10^{-2}$ torr. The evaporation of refractories under such conditions is nevertheless rather difficult. Conventional electron beams cannot be used in the higher vapor pressure environments of the liquid containing vessels due to plasma discharges and severe degradation of organic materials. By circulating liquid nitrogen through a cooling stage to cryopump the hearth area, an electron gun can be used even when high ambient vapor pressures ($>10^{-5}$ torr) exist in the reaction flask. This is a significant adaptation of the electron gun, since it is otherwise not possible to conduct electron beam experiments involving volatile liquid solutions in a rotary reactor of this kind. The liquid nitrogen (or water) cooling to the source assembly can be seen in Figures 2 and 4. It consists of a stainless-steel, thin-walled vacuum feedthrough unit having an o-ring seal to the main flange assembly. The coolant is channelled through two adjustable length (1/4 in dia.) stainless-steel tubes to the copper block depicted in the close-up in Figure 2. The copper block serves (a) to support the electrostatic focusing shields, (b) to house the quartz crystal microbalance, and (c) to cryopump the region in the immediate vicinity of the hearths. Additional cryopumping is obtained by attaching stainless-steel foil to the top and sides of the copper cooling block, tracing backwards a line approximately 10 inches. This latter shield also protects the electrical feedthrough insulation from metal contamination, thereby reducing the possibility of electrical short circuit. Although the electron beam focus is disturbed, the cryopumping capacity of the gun can be further enhanced by expanding the 77K surface area around the source zone. This is done by wrapping a 10 in. x 3 in. stainless-steel foil about the hearth area, affixing each end to opposite sides of the liquid nitrogen cooled block. A carbon liner, 2 mm thick and tightly fitted to the copper hearth (Figure 6) markedly improves yields of evaporate and results in fewer interruptions of the electron beam evaporation.(13) In short, we achieve higher evaporation rates at reduced accelerating potentials, for roughly fixed emission current. The combined effect is to reduce the frequency of discharges and short-circuiting of the gun and to decrease the level of soft x-ray radiation.

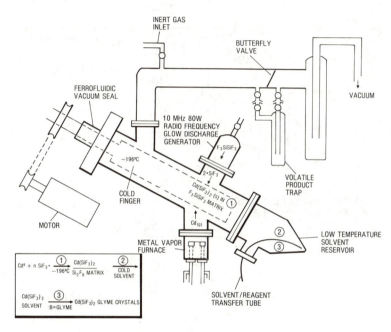

Figure 5a.   Rotatable cold finger (after Lagow et al. [29]).

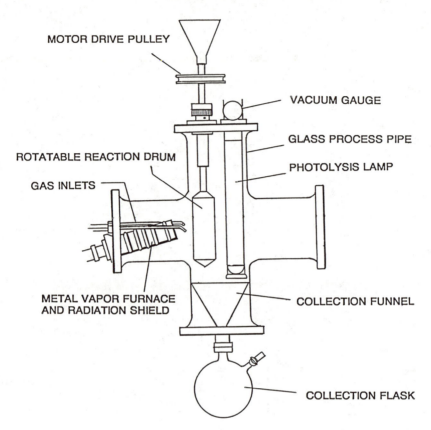

Figure 5b. Rotatable cold finger photochemical metal atom reactor (after Billups et al. [30]).

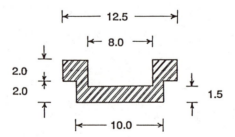

CARBON HEARTH LINER: CROSS SECTION

(Dimensions in mm)

Figure 6. Graphite crucible for lining water-cooled copper hearths of reverse polarity electron beam furnaces.

Microscopics to Macroscopics

Chemistry from spectroscopy is one way to view a technique for elic-
iting information about the effects of deposition rates, thermal
stability, photochemistry, reactivity of products towards secondary
reagents, criteria for chemical work-up -- in short, for feasibility
studies of vapor synthesis reactions. Although often stubborn in
complexity, cryogenic matrix isolation synthesis and spectoscopy can
provide important insights of this kind. The static reactor
described by Ozin(21) for synthesis at sub-20 K temperatures can be
viewed as a more or less direct scale-up of the microscopic experiment. An appreciation of the microscopic processes involving mass
transport at the condensation step and in the mobilization of reactive species on warm-up can be as important to a successful experiment as a good synthetic strategy. (An insightful analysis of the
cocondensation experiment is given in Chaper 2 of the book by
Blackboro and Young.(6)) The use of VS to make new organometallic
compounds capable of depositing metallic films at moderate temperatures (<100° C) has attracted some interest.(31) Matrix isolation
spectroscopy has identified a number of Ni-olefin complexes,
$Ni_n(olefin)_x$ (n=1,2,3; x=1,2,...), formed by controlled Ni atom
aggregation in olefin-doped rare gas solids.(32) Moreover, studies
of these and other homoleptic nickel-olefin complexes, indicate that
certain systems might be stable enough for macroscale slynthesis.
Berry(31) has translated the suggestions of the matrix isolation
work into a project to form nickel thin films from labile Ni-olefin
complexes. His experiments revealed that the homoleptic fluoro- or
chloro-complexes were too unstable to be isolated by standard VS
techniques; however, the fact that isolable adducts with PF_2 did
deposit films between -16 and +6°C is encouraging.
     Similarly, the design of macroscale vapor synthesis reactions in
liquids can benefit from adaptations of the techniques of cryogenic
matrix isolation for studies in fluids.(33,34) For example, a significant development in the field of VS has been the discovery by
Klabunde and co-workers(35) of "solvated" metal atoms for use in the
preparation of uni-and bimetallic supported catalysts. It is worth
examining the subject of solvated atoms in a little detail to reveal
something about them and to introduce the microscale VS technique in
thin, quiescent liquid films.
     Toluene is a common "solvating" agent in these static reactor
cocondensation reactions. It is thought that the molecule stabi-
lizes metal atoms against catastrophic aggregation at 77 K by forming weak π-complexes with elements like Fe, Co, Ni, Pd, Pt and Ru.
The toluene complex of Ni is a source of Ni(0), and ligands can be
added (triphenylphosphine, triethylphosphite, CS_2) which will dis-
place the toluene molecules. The thermal stability of the complexes
increases with a more electron demanding ligand like ε,α,α trifluo-
romethylbenzene. Information clarifying the nature of the arene-
metal atom interaction, as well as the stages of aggregation to give
metal microcrystallites, might assist in the design of VS experiments to impart control over particle size distributions, or to
optimize yields of a particular metal particle coomposition (as in
the preparation of alloyed particles). Briefly, Klabunde envisages
the following stages of Ni atom complexation and aggregation:

$$Ni_{(g)} + toluene \xrightarrow{77K} M\text{-solvent complex} \xrightarrow{meltdown}$$
$$(colored)$$
$$(I)$$

$$M\text{-solution} \xrightarrow[\text{warming}]{\text{further}} M_n\text{-solvent slurry} \xrightarrow[25°C]{\text{vaporize solvent}}$$
$$II \qquad\qquad (black)$$
$$III$$

$$M_n\text{-solvent adduct} \xrightarrow{pyrolysis} M_n + organics$$
$$V$$

$$(black; small$$
$$crystallites)$$
$$IV$$

To model the stages in Klabunde's scheme, a solid and liquid phase analog of the macroscale VS experiment can be used. Reference to Figure 7 will help in understanding procedure. The Figure shows a cross-sectional view of a type of vacuum furnace/closed-cycle helium refrigeration system conventionally used for cryogenic matrix isola-tion synthesis,(5,7) but here modified for vapor depositions into cooled liquid films.(33) To probe the stages of the scheme, a miniature cocondensation experiment is conducted. The appropriate optical window (NaCl or Suprasil for uv-vis studies) is sandwiched between indium gaskets within a copper housing and held horizontally so that a liquid film of uniform thickness can be maintained when the solid matrix is subsequently melted (Stage II) and warmed (Stage III). The substrate window assembly is attached to the second stage of a Displex closed-cycle helium refrigerator. The vacuum furnace can be rotated into any one of the positions A, B or C. Orientatin B shows the furnace directly above the optical window. Monatomic nickel is generated by resistance evaporation from a nickel filament (28x5x0.25 mm, 99.999%). The geometry-corrected(33,34) downward deposition rate is monitored by a quartz crystal microbalance. Atoms are generally deposited at 10-5000 ng-s^{-1}, ie, under condi-tions favoring Ni-complex formation. A vapor of toluene and meth-ylcyclohexane 1:10 v/v (methylcyclohexane is a good glass former) is metered in through the ports marked E and condensed at 77 K with Ni. After deposition, the furnace is moved to position C where the beam from a spectrometer can interrogate the sample. Ultraviolet-visible studies of matrices prepared in this manner led to the conclusion that the "solvated nickel atom" probably emerges as a bona fide complex from a Ni/toluene static reactor cocondensation. It is also likely that very low nuclearity Ni clusters coexist with the complex at 77 K (Stage I). The second stage is reached on warming the matrix to 120 K or melting at 150 K, when nickel clusters of colloi-dal dimensions emerge. Additional warming to 170-180 K and then solvent evaporation at 200-290 K draw the sample through Stages III and IV where the aggregates display optical properties like those of nickel microcrystallites.

Solvated metal atoms and aggregates can also be made by liquid phase rotary reaction techniques.(36-39) Depending on the reactivi-ty of the medium, clustering of metal atoms may occur immediately

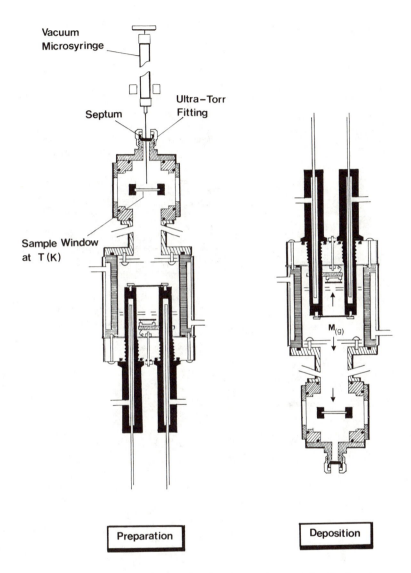

Figure 7. Conversion of standard cryogenic matrix isolation equipment for use in microscale, liquid phase metal vapor synthesis and spectroscopy. Continued on next page.

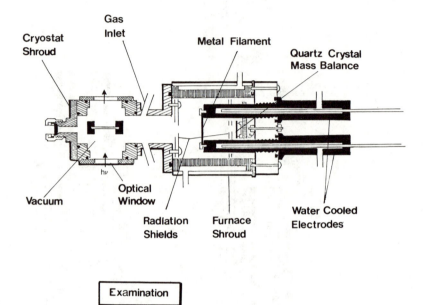

Examination

Figure 7. <u>Continued</u>. Conversion of standard cryogenic matrix
isolation equipment for use in microscale, liquid phase metal
vapor synthesis and spectroscopy.

under diffusion control, or it may be prevented entirely by capturing the atom as a labile organometallic complex as in the cases of the bis(arene)Fe or Ru molecules. Here too the utility of the microsolution technique for probing reactivity in the liquid phase is demonstrated. In orientation A in Figure 7 we show a vacuum feedthrough consisting of a gas chromatographic septum sealed within a 0.25 in. Cajon Ultra Torr fitting, half of which has been silver-soldered to a square flange attached to the base of the cryostat shroud. Liquids having a low vapor pressure when cooled sufficiently (but not easily frozen) can be injected by means of a vacuum microsyringe (Pressure-Lok series "A-2" 100 μL syringe, Dynatech Precision Sampling Corp., Baton Rouge, La.) onto the pre-cooled optical window. The injected liquid may, for example, contain a ligand targeted for metal atom reaction.(36,37) Because the fate of deposited diffusing metal atoms can be traced quantitatively in this apparatus, important information can be obtained regarding effects of metal atom deposition rate, substrate temperature and viscosity, reactivity towards secondary ligands, the kinetics of product formation, product distributions, metal atom diffusion-aggregation processes, thermal and photostability. Such applications are illustrated in a series of papers dealing with microscale metal atom chemistry of arene and naphthyl-functionalized oligomers and polymers in the liquid phase.(37,40)

Microsolution studies(33,34,36,37) give some insights into the growth and stabilization of colloidal particles produced by aggregation of atoms condensed into an organic medium. Silver atoms deposited into weakly interacting liquids such as polyolefins, olefin oligomers, low molecular weight ethers and crown ethers, agglomerate to particles (1.0 nm<R<20 nm) showing a size distribution which remains independent of silver loading at a fixed temperature during the deposition period.(36) The effect manifests itself optically in the linear growth of the surface plasmon absorption with increasing amount of deposited Ag, and in the invariance of the peak position of maximum absorption and the half-bandwidth. Particles isolated in the oligomers and polymers showed greater stability towards thermally induced aggregation to larger particles than did those hosted in the low molecular weight ether solvents. From cocondensation matrix isolation experiments(41) it is also clear that the isolation of silver atoms and aggregates and their mobilization to give larger particles are very sensitive functions of the host and the isolation temperature. This information on colloid formation in the liquid media has been found useful in the macroscale preparation of Ag- and Pd-carbon composites for use in electrocatalysis.(36) Similar insights are possible for colloidal metal systems derived from the melts of static reactor cocondensation reactions.(42-45)

A last example of microscale metal atom chemistry makes an important link with the macroscale experiment. A low temperature form of a complex between iron and toluene has been identified(34) in cocondensates formed at 77K. Controlled warming converts the product to a high temperature species, stable in the liquified complex/toluene/methylcyclohexane solution. Apparently, the same high temperature complex can be formed directly by iron atom titrations of a 10% toluene/methylcyclohexane solution at 150 K. The high temperature form begins depositing colloidal iron above -30°C.

The iron vapor-toluene reaction has evoked interest because of the
lability of the proposed bis(arene)iron complex to ligand subsitu-
tion and to loss of both toluene molecules to free the metal atom.
In the latter case the toluene molecules may be usefully regarded as
metal atom carriers which can be used to direct the latent reactivity of the atom in subsequent solution phase chemistry. In this way
the metal atom experiment can benefit from the convenience and
additional versatility afforded by bench-top chemical manipulations.
These results are relevant to a reported preparation of a dehydroxylated silica-supported Fischer-Tropsch catalyst from a static
reactor codeposition of Fe and toluene.(46) In the liquid phase,
iron atoms "bottled" in this way have also been utilized in an
exceedingly mild method for making minute catalytically active
superparamagnetic clusters on the surface and within the cavities of
a dehydrated sodium zeolite Y.(38) Using the rotary reactor, pre-
formed solutions of solvated iron atoms (as the toluene complex) are
cannulated below their decomposition temperature out of the flask to
a cold slurry of the support in toluene. Diffusion of intact
complex to the zeolite surface and within the cavities is followed
by cold solvent washes and filtration. The metal/support
combination is warmed to room temperature under dynamic vacuum,
causing desolvation, and metal atom aggregation on the surface and
within the support. Besides their chemical activity, iron clusters
housed within the cavities of zeolites may show interesting
collective magnetic properties such as the antiferromagnetic
ordering of superparamagnetic moments detected by Schmidt.(47) This
physical property leads us to the next section of this chapter which
addresses some possible applications of vapor synthesis in materials
science.

Applications

For metastable structure synthesis, the VS technique can be added to
the low pressure varieties--laser pulsing, gas aggregation, ion
implantation, ion beam mixing--with ultrahigh rates of up or down
quenching. In principle it should be possible, by condensation onto
substrates held at temperatures below those of configurational
freezing, to form homophase solids of arbitrary composition.
Structures with exceptionally high interfacial area densities should
be accessible. In this category we include ultrafine interphase
dispersions with high interphase boundary energy and interface to
bulk volume ratios. Some of these dispersions are unique for the
extremely high density (10^{13} to 10^{14} cm-cm^{-3}) of dislocatins which
can be trapped in one of the phases. Novel compositionally and/or
topologically disordered structures may be produced.
     Nanocrystalline materials comprising sub-100 l metal particles,
when compressed to 50% of their bulk density, show properties
(specific heat, thermal conductivity, saturation magnetization and
critical temperature for superconductivity) provocatively different
from those of their crystalline or glassy counterparts.(48) It is
well known that the interfaces of mechanically reduced composites
are effective in interacting with dislocations and with flux lines
in superconducting composites. Precursor materials for the preparation of ultrafine filamentary composites can also be imagined. Here
the combinations of interphasial boundaries and dislocations can

dramatically alter the electron and phonon characteristics of materials, resulting in properties which come into play over the same
microstructural scale; viz., materials having unusual magnetic
properties in addition to having strength and good normal-state
conductivity.

The development of chemical techniques associated with vapor
synthesis for manipulating atoms to give mondispersions of simple
geometry particles (spheres, spheroids), would contribute substantially to a number of areas. For example, magnetically anisometric
particles of defined size, axial ratio and crystallinity are required by the best recording devices. Moreover, if spherical, uniform
diameter inorganic pigment particles can be made, then exact theoretical calculations of their optical properties can be useful in
directing control of color and purity. Medical applications include
the design of drug delivery systems in which the active component is
either adsorbed on fine particles or encapsulated, alone or with a
magnetic colloid, in a particle covered by a semipermeable membrane.

Ultrastructure processing of materials of predetermined properties is desireable in ceramic fabrication. Of importance here is
the synthesis of materials having tailored chemical, structural and
morphological characteristics. Proto-materials in the form of specialty chemical precursors such as mono-and multi-metallic alkoxides
might be prepared by direct reaction of metal atoms and alcohols.
These may be hydrolysed in a controlled manner to inorganic colloidal sols of uniform size distribution. Timms has envisaged the
synthesis of new inorganic coordination compounds via direct reaction of atoms and ligands.(2) The products of such reactions may be
channeled into reducing or oxidizing environments to produce a host
of morphologically distinct metal oxides. Oxophilic metal atoms may
be directed to oxygen atom sources such as epoxides and N-oxides
under the intrinsically low temperature and pressure conditions of a
VS reactor.(49) Reductive deoxygenation of the organic molecules
may give new oxides directly. Most metal oxides dissociate(7,8)
when they evaporate, however, yielding the metal atom or a lower
oxide of the metal. Based on studies of the reactivity of moly-
bdenum and tungsten trioxide vapors towards organic and inorganic
compounds,(50,51) it is reasonable to expect that the vapors of
molecular high temperature species might be reacted to yield unconventional metal oxide compounds. Unique combinations may result
simply because in some cases the products of reaction will be kinetic in origin. Moreover, adaptations of the VS apparatus, hand in
hand with chemical techniques hybridized to deal with the products
of VS and the demands for microstructural control in composites,
might be scaled to examine relatively unexplored areas in composite
electroceramics. Such areas include single domain behavior in
ferroelectric particles, and dielectric phenmomena found in relaxor
ferroelectrics (fluctuating microdomains in the superparaelectric
state) which resemble those found in Neel superparamagnets.

Where VS can contribute to the synthesis of new materials, it
can clearly also contribute, by implication, to fundamental studies
of matter. For example, through the synthesis of well-defined colloids and metal clusters, research in adsorption, adhesion, mineral
processing, corrosion, pollution and the physics of spatially con-
fined structures is enhanced.

A metal atom reactor is often used in a variety of "dirty" and "clean" operations. Accordingly, it should be remembered that radiation, adventitious water, oxygen, hydrocarbons, metallic particles and so on, can affect the properties of the isolated products. In the case of minute magnetic structures it is important to determine clearly the role of such agents in affecting volume and surface magnetic properties. Incorporation of a high vacuum Schlenk manifold such as the one described by Wayda in this book, should be considered an important supplement to the VS equipment.

Compared with the established technologies and techniques of materials science, vapor synthesis has really yet to declare itself. Apart from catalysis related applications, there have been few direct assaults on new materials from VS. A brief suggestive sketch is given below where some of the early manifestations of this potential can been seen.

Magnetic Materials from Metal Atom Reactions.

The first reports of fine powders of magnetic nickel particles produced by VS emerged from laboratories at the University of North Dakota and IBM.(52,53) Klabunde and co-workers(52) showed that nickel crystallites (mean diameter from x-ray diffraction <35 Å) form when the atoms cluster at low temperatures in alkane media. The particles are ferromagnetic or nonferromagnetic depending on the preparative conditions. Large amounts of carbonaceous material (as C,H but not Ni carbides) were incorporated in the crystallites due to alkane cleavage at low temperatures (<138 K). Scott et al.(53) carried out a more detailed examination of nickel microcrystallites synthesized by warm-up of static reactor cocondensates of Ni/toluene, Ni/SF_6 or Ni/Xe. Transmission electron microscopy (TEM) revealed particle diameters less than 16 Å for nickel microcrystallites derived from toluene. Crystallites harvested from SF_6 or Xe were larger in size (35-50 Å and >100 Å, respectively), and incorporated fewer impurities. Up to 10% of the total sample weight from the toluene derivative included organic material thought to be incorporated as a thin amorphous layer about the particle surface. The largest ferromagnetic particles showed magnetic moments of about 40% of the bulk nickel value, whereas paramagnetism ws observed for the smallest metallic clusters. The latter showed a Curie-like susceptibility with an exchange enhanced moment. The aggregation processes leading to such minute particles may be self-limiting in the organic media.(53) Rapid accumulation of organic fragments during melt-down and warm-up may prevent further accretion of metal atoms once the nuclei have reached a critical size. This suggests that it may be possible to manipulate particle sizes by controlling the reaction chemistry in the microenvironment where nucleation takes place. A prescription for doing this is revealed in the work by Timms and collaborators.(54) Like iron, cobalt combines with toluene giving a thermally unstable, formally zerovalent mononuclear metal complex. The molecule decomposes at around -70°C to liberate the atom. The Timms approach combines liquid phase metal atom synthesis with the classical techniques of colloid chemistry for producting and controlling the dimensions of metallic and semiconducting materials, namely the use of micelles to mediate reaction

chemistry and to confine particles to narrow size distributions.(55)
In essence cobalt atoms are deposited into a 150 K toluene/manoxol-
OT liquid in the rotating reactor, where they probably form the
delicate bis(toluene)Co molecule, possibly within the local environment of the surfactant. Decomposition to cobalt atoms, followed
by sub-ambient temperature aggregation presumably within a protective micellar sheath, yields approximately spherical particles with
a mean diameter (TEM) of 50 ≠ 6Å (cf. organometallic routes which
typically produce 70 ± 10Å particles(56)). The superparamagnetic
fluid gave a saturation magnetization of 10.5 $JT^{-1}kg^{-1}$. Broad
compositional ranges might be achieved for alloy systems by suitable
adaptations of the technique.

Colloids and Catalysts from Metal Atom Reactions.

Timms' ferromagnetic fluid is truly a colloidal dispersion. Chemically unreactive liquid polydimethylsiloxanes are attractive as
colloid dispersants, and Francis and Timms(57) have investigated
depositions of Cr, Fe, Co and Ni atoms into a Dow Corning fluid
(DC200) spun at 220-250 K in a rotating reactor. Dispersed metal
colloids result. The phenyl-substituted analog of DC200 has the
ability to trap Fe atoms between the pendent rings when reactions
are conducted at sufficiently low temperatures. The phenyl substi-
tuents can therefore interfere with the re-polymerizatin of the
metal atoms which find themselves encapsulated within the siloxane
polymers. Gradual warming of the fluid causes the arenes to expel
the atoms, which then aggregate within the fluid. A thermally more
robust bis(cyclopentadiene)Fe(0) complex tethered to a siloxane
backbone has been observed to react with O_2 giving an iron oxide
entrained within the polymer.(58) The magnetic properties of these
siloxane polymer-colloidal dispersions have not yet been determined,
although the origin of these materials in vapor synthesis opens
possibilities for the development of interesting polymer composites.
Particles which can be deposited as metal films by stripping the
organic diluent are termed "living colloids".(42,43) These are made
by cocondensing the desired element with an organic compound at 77 K
and warming the mixture to room temperature. Films produced by
stripping acetone from Pd or Sn colloids act as semiconductors,
although it is not clear what is responsible for the behavior.
Colloids result when silver atoms are deposited at controlled rates
into various cooled organic liquids.(36) Some aspects of their
growth patterns and stability towards further agglomeration were
described earlier under uses of the microsolution experiment. The
procedure for making colloids in a liquid phase rotary reactor is
the subject of a process patent.(59) The approach, which is conceptually similar to the solvated metal atom dispersion technique of
Klabunde, is likely to generate interest in any application requiring ultrafinely dispersed, supported metal particles. Ag/C and Pd/C
compositions yielded in this way, have been fashioned into oxygen
gas porous electrodes for use in an alkaline H_2/O_2 fuel cell. The
largest body of information on catalysis-related uses of VS has come
from the laboratory of Klabunde.(7,35,39,46,60) Work ranging from
the near-explosive polymerization of butadiene(60) to Fischer-
Tropsch catalysis(46) has demonstrated, where comparisons can be

made, that VS is capable of producing materials showing selectivity and activity significantly improved over conventionally produced supported catalysts. Pivotal has been the synthesis of supported "early-late" bimetallic transition metal catalysts.(61) Work-up of these metal vapor/toluene cocondensation reactions, gives "pseudoorganometallic" powders which are claimed to be amorphous. Alkenes in contact with H_2 and Mn-Co pseudoorganometallic powders undergo hydrogenation at -60°C at diffusion controlled rates. It is instructive to note that much of the groundwork here has been established using some of the simplest of metal atom reactor designs. Clearly, significant contributions can be made using apparatuses which are less expensive and less complicated than some of those described above.

A patented process for transporting iron and cobalt atoms to the surfaces and interstitial regions of zeolites has been described. (38) The compositions are Fischer-Tropsch catalysts. Iron catalysts figure prominantly in ammonia synthesis. Aqueous Fe_n/TiO_2 colloids may be prepared by decomposing bis(toluene)Fe on TiO_2.(62) The resulting composites photoconvert N_2 to ammonia. Vapor synthesis has been used to prepare materials which will oligomerize or polymerize olefins. For example, since a nickel atom is the only continuing member of the cycle in which butadiene is cyclotrimerized, the process can be viewed as catalysis by an atom. Simple retrosynthesis of tris(butadiene)Ni suggests that it ought to be accessible by VS, and indeed this is the case.(63) This finding is part of a wider study which used the paradigm of Ziegler-Natta catalysis to explore VS-based procedures for rapid screening of oligomerization and polymerization catalysts and co-catalysts.(63) High defect $MgCl_2$ structures suitable as $MgCl_2-$ supported Ziegler-Natta catalysts can be isolated from cocondensation reactions of $MgCl_2$ vapors, a co-dispersant and $TiCl_4$.(64) Additional applications of VS to polymer science can be found in a review of the field,(40) where six important areas of synthesis have been identified where VS has contributed. These are (i) the synthesis of organometallic monomers; (ii) monomer polymerization induced by high temperature vapors; (iii) preparation of organometallic polymers by direct reaction of polymers with metal atoms; (iv) the concatenation of organic molecules by metal atoms to form oligomers; (v) the use of polymers to host metal atom aggregation; and (vi) the preparation and screening of polymerization catalysts. The following section describes a non-catalytic use of metal atoms for polymerizing monomers to an important class of materials--block copolymers.

A Macroscopic Application: Formation of Block Copolymers by Sodium Atom Induced Anionic Polymerization

Like certain combinations of metallic elements which show degrees of mutual solubility, so organic polymer blocks show varying tendencies to "alloy" in the solid state. The synthesis and properties of block polymers is a developing science which leads to important materials and engineering applications.(65) In this section we describe the preparation of a BAB block terpolymer of ⍺-methylstyrene and styrene.(66) Sodium atoms are used to initiate polymerization in liquid tetrahydrofuran solutions of monomer cooled

to 145 in a rotating reactor.   The experiment illustrates (i) a
grease-free ligand inlet system; (ii) homogeneous liquid phase metal
atom induced polymerization; (ii) low temperature cannulation; (iv)
in situ multi-step synthesis.   (CAUTION: There is always a danger
of explosion/implosion in vapor synthesis reactions.)

The rotating reactor used in this preparation was custom built
by Torrovap Industries Inc.(13,16)   Figure 8 illustrates the experimental set-up schematically.   Monomers are purified as described
elsewhere.(66)   Tetrahydrofuran is distilled from a solution of
living polystyrene formed by combining equivalent volumes of monomer
and n-butyllithium until a distinctive red solution was obtained.
THF distilled from sodium/benzophenone proved to contain residual
ketone which reacted with Na atoms in the reactor and very effectively quenched the polymerization.

Prior use the 5 L Pyrex reaction vessel is cleaned (conc.
$H_2SO_4/30\%H_2O_2$), rinsed several times with water and baked at 180°C.
Three small sodium pellets (400 mg, Aldrich) are placed in a 1 cm x
1 cm quartz crucible situated in a triple-strand tungsten wire
spiral basket (R. D. Mathis Co., Long Beach, Ca.).   Provided the
crucible is not heated too quickly, it is not necessary to pre-melt
the sodium since vaporization of sufficient quantities of atoms
occurs below the temperature at which the material boils and spat-
ters.   Typically, the hot flask is mounted over the electrodes,
bolted to the rotary flange and quickly evacuated to 5×10^{-6} torr in
about 20 minutes, and to 2×10^{-7} torr after several hours.

A manifold for introducing three or more liquid reagents is
diagrammed in Figure 9.   The tree comprises greaseless conical
joints manufactured by J. Young, Ltd.(67)   The size B19 joints consist of a polished glass cone fitted with a P.T.F.E.-over-Viton o-
ring which makes a high vacuum seal against a polished glass socket.
The tapered joints incorporate drip tubes to reduce contact of solvents with the joints.   These joints are isolated from the vacuum
system by J. Young low hold-up, vacuum P.T.F.E. taps (item SPOR/8, 8
mm bore).   A tap at the rear of the tree (not shown in the Figure)
allows it to be connected to an auxiliary Schlenk line for separate
pumping of the dead volumes or for inert gas flow.   The base of the
manifold inserts into a greaseless spherical o-ring ball joint which
communicates with a polished glass socket (J. Young GBM/18/9 (ball),
GBF/18/9 (socket); 9 mm shank bore, 12 mm o.d.).   A viton o-ring
provides a better vacuum seal here than does one which is P.T.F.E.
coated.   A drip tube at this junction lengthens the service life of
this o-ring.   The socket joint tapers to a Kovar seal welded to a
1/8 in. o.d. stainless-steel cannula.   The ball-and-socket joint
permits a strain free connection of the ligand inlet manifold to the
vacuum system.   We have inserted a heated micrometering valve
(Whitey, SS-22RS4) between the ligand manifold and the vacuum as a
matter of convenience for solvent flow control and vacuum shut-off.
However, a valve with a solvent-inert packing is desireable.   The
smallest valves having finest flow control must be meticulously
maintained to avoid accumulations of impurities which can be detrimental to these kinds of reactions.   Viton o-rings can be replaced
with Kalrez.

When the system has achieved a vacuum $<5 \times 10^{-5}$ torr, liquid
nitrogen is added to the trap and a pentane slush bath is elevated

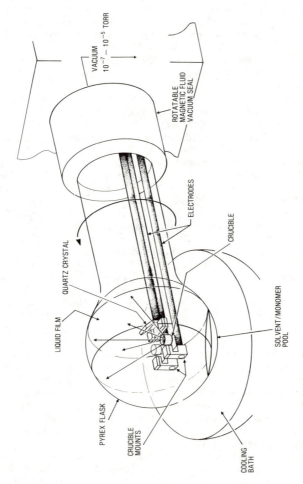

Figure 8. Rotatable metal atom reactor for conducting liquid phase metal atom induced polymerizations of organic monomers.

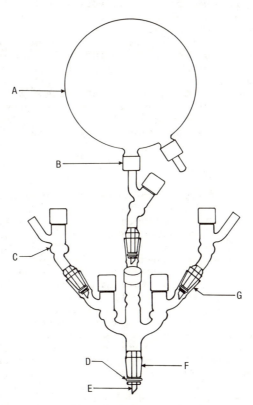

Figure 9.   Greaseless ligand inlet manifold.   A: Round bottom flask; B: Ace Thread Nylon bushing with Kalrez o-ring seal;   C: J. Young P.T.F.E. low hold-up high vacuum 8 mm bore stopcock; D: Teflon covered o-ring;   E: Drip tube; F: J. Young polished glass B19 cone fitted with an o-ring groove; G: J. Young polished glass B19 socket.

about the reactor flask inclined at 5° and rotating at about 60 rpm. A 10% v/v solution of degassed α-methylstyrene is introduced to the flask at such a rate that the pressure does not exceed 10^{-5} torr. Complete addition of the fluid is indicated by a drop in pressure to $1-3\times10^{-5}$ torr. Freezing of the solution is avoided by adjusting the speed of rotation, the height and temperature of the cooling bath. The quartz crucible is brought to 13 A over a period of 15 minutes. The last stages of crucible heating must be done carefully to avoid sudden evaporation of excessive sodium. The current required to supply a sufficient amount of atoms for initiation is very reproducible for our configuration. (The quartz crystal mass monitor is not sensitive enough to quantify this light element and the very short deposition time.) Vaporization is stopped abruptly upon the first sign of initiation (reddening of the solution). Polymerization is allowed to proceed in the spinning flask, under vacuum at about 150K.

Low Temperature Cannulation.

The homopolymer is sampled to determine the block length. The diffusion pump and cold trap are isolated from the reaction flask by a butterfly valve. Dry, oxygen-free argon is metered in to equilibrate the vessel at atmospheric pressure. Nitrogen gas is circulated through about 250 cm of 1/4 in. o.d. coiled copper tubing immersed in liquid nitrogen, and then through a jacketed cannula equipped with a thermocouple (Fe/constantan) positioned near the sample inside the flask. Lower temperatures are achieved at faster gas flow rates. The cannula ends in a teflon siphon which can be positioned by a mechanical manipulator from the atmospheric side of the reactor. A suitable receiver (Schlenk tube) is attached to the transfer tube by means of a custom adaptor. We have fashioned a glass piston tap adapter from a J. Young 15 mm bore stopcock. Connection to an o-ring fitted, brass ball joint at the end of the transfer tube is made to a compatible GBF/28/15 socket joint in the adapter.

The dry receiver is evacuated through the Schlenk line and then cooled to 150 K in pentane slush. The siphon is dipped into the living polymer solution and the glass piston in the adapter withdrawn to allow a sample to collect in the receiver flask. The tap is quickly closed and the diffusion pump is isolated completely. The butterfly valve is opened to the rough pump to re-establish vacuum.(CAUTION: Ar will condense in the cold trap.) When diffusion pumping has brought the pressure down to $<3\times10^{-5}$ torr, styrene monomer can be metered into the rotating flask. The blood red solution will not decolorize if initiation of the styrene monomer takes place. The molecular weight of the styrene component in the final product can be controlled through the amount of added monomer. The living polymers are quenched by adding trace amounts of methanol or the desired chain terminator.

Closing Remarks

Future areas of application of vapor synthesis are easily pinpointed by extrapolating its origins in thin film science and vacuum technology to those regions of solid state chemistry and physics where paradigms for chemical synthesis at the atomic or molecular level can be discovered. For example, chemical approaches to the following might originate in vapor synthesis: the preparation of clusters suitable for studies of small particle physics; the low temperature synthesis of mono-and multi-metallic oxides, nitrides, silicides and carbides; reactions involving layered structures; the preparation of super-, superionic and low dimensional conductors, alloys, amorphous solids and magnetic structures. The virtue of this particular technique in perhaps contributing uniquely to these areas and others, might be seen to reside in an intrinsically molecular approach to materials synthesis with compositional and microstructural control determined jointly by technology and the techniques of chemistry at the molecular level; and in the qualitative and quantitative ways the technique differs from others which utilize atomic and molecular beams and discharges.

Acknowledgments

It is a pleasure to acknowledge helpful discussions with Mr. V. E. Lamberti, Dr. A. L. Wayda, and Professors G. A. Ozin and K. J. Klabunde. The technique for alkali metal atom induced anionic polymerization was developed in collaboration with Ms. S. A. Heffner and Dr. M. E. Galvin. Thanks are due to Torrovap Industries and G. V. Planar Ltd. for providing technical information about their metal atom reactors.

Literature Cited

1. Skell, P. S.; Wescott, L. D.; Goldstein, J. P.; Engel, R. J. Am. Chem. Soc. 1965, 87, 2829.
2. Timms, P. L. Proc. R. Soc. London Ser. A 1984, 396, 1.
3. Timms, P. L. Turney, T. W. Adv. Organomet. Chem. 1977, 15, 53.
4. Timms, P. L. Adv. Inorg. Chem. Radiochem. 1972, 14, 121.
5. Moskovits, M.; Ozin, G. A., Eds. Cryochemistry; John Wiley and Sons, Inc.: New York, 1976.
6. Blackboro, J. R.; Young, D. Metal Vapor Synthesis in Organometallic Chemistry; Springer-Verlag: Berlin, 1979.
7. Klabunde, K. J. Chemistry of Free atoms and Particles; Academic Press: New York, 1980.
8. Klabunde, K. J., Ed. Thin Films from Free Atoms and Particles; Academic Press: New York, 1985.
9. Metal Atoms in Chemical Synthesis, Symposium sponsored by Merck'sche Gesellschaft fur Kunstk and Wissenschaft e. V. at Darmstadt, FRG, May 12-15, 1974, as reported in Angew. Chem. Int. Ed. Engl. 1975, 14(4), 193 and (5), 273.
10. Klabunde, K. J. In Reactive Intermediates; Abramovitch, R. A., Plenum Press: New York, 1980; vol. 1, p. 37.
11. Ozin, G. A.; Power, W. J. Adv. Inorg. Chem. Radiochem 1980, 23, 79.
12. Ozin, G. A. Chemtech Aug. 1985, 15, 488.
13. Andrews, M. P.; Ozin, G. A.; Francis, C. G.; Huber, H. X.; Molnar, K. Inorg. Chem. in press.
14. Green, M. L. H. J. Organomet. Chem. 1980, 200, 119.

15. Klabunde, K. J.; Groshens, T.; Efner, H. F.; Kramer, M. J. Organomet. Chem. 1978, 157, 91.
16. Torrovap Industries, Inc., 90 Nolan Court, Unit 39-40, Markham, Ontario, L3R 4L9, Canada.
17. G. V. Planar, Ltd., Windmill Rd., Sunbury-On-Thames, Middlesex, TW 16 7HD, England.
18. This circuit is similar to those utilized in units manufactured by Torrovap Industries.
19. Unvala, B. A. Le Vide 1963, 104, 109.
20. A cryopump option is offered by the makers of the Planar apparatus.
21. Godber, J.; Huber, H. X.; Ozin, G. A. Inorg. Chem. 1986, 25, 2909.
22. Howard, J. A.; Sutcliffe, R.; Mile, B. J. Catal. 1984, 90, 156.
23. By dissolving ethylene in a flow of liquid methylcyclohexane the volatile olefin can be "gettered" and condensed on the walls of a 77 K rotating reactor.
24. Stockbridge, C. D. In Vacuum Microbalance Techniques; Behrndt, K. H., Ed.; Plenum Press: New York, 1966; vol. 5.
25. Eschbach, H. L.; Krindhof, E. W. In Vacuum Microbalance Techniques; Behrndt, K. H. Ed.; Plenum Press: New York, 1966; vol. 5.
26. Lu, C. S. J. Vac. Sci. Technology 1975, 12, 578.
27. Kontes Scientific Glassware/Instruments, Spruce St., Box 729 Vineland New Jersey, USA.
28. Bennett, J. E.; Thomas, A. Proc. Roy. Soc. London Ser. A 1961, 128, 123.
29. Guerra, M. A.; Bierschenk, T. R.; Lagow, R. J. submitted for publication.
30. Billups, W. E., Bell, J. P.; Hauge, R. H., Kline, E. S.; Moorehead, A. W.; Margrave, J. L.; McCormick, F. B. Organometallics 1986, 5, 1917.
31. Berry, A. D. Organometallics 1983, 3, 895.
32. Ozin, G. A.; Power, W. J.; Upton, T. W.; Goddard, W. A., III J. Am. Chem. Soc. 1980, 100, 4750.
33. Ozin, G. A.; Francis, C. G.; Huber, H. X.; Andrews, M.; Nazar, L. J. Am. Chem. Soc. 1981, 103, 2453.
34. Ozin, G. A.; Francis, C. G.; Huber, H. X.; Nazar, L. Inorg. Chem. 1981, 20, 3635.
35. Klabunde, K. J.; Efner, H. F.; Murdock, T. O.; Ropple, R. J. Am. Chem. Soc. 1976, 98, 1021; Klabnde, K. J.; Davis, S. C. ibid, 1978, 100, 5973; Klabunde, K. J.; Davis, S. C.; Hattori, H.; Tanaka, Y. J. Catal. 1978, 54, 254; Klabunde, K. J.; Ralston, D.; Zoellner, R.; Hattori, H.; Tanaka, Y. J. Catal. 1978, 55, 213; Davis, S. C.; Klabulnde, K. J. Chem. Rev. 1982, 82, 152.
36. Andrews, M. P.; Ozin, G. A. J. Phys. Chem. 1986, 90, 2929.
37. Ozin, G. A.: Andrews, M. P. In Studies in Surface Science and Catalysis; Gates, B. C.; Guczi, L.; Knozinger, H., Eds.; Elsevier; New York: 1986; vol. 29, p. 265.
38. Nazar, L. F.; Ozin, G. A.; Hugues, F.; Godber, J.; Rancourt, D. J. Mol. Catal. 1983, 21, 313.
39. Klabunde, K. J.; Tanaka, Y. J. Mol. Catal. 1983, 21, 57.
40. Andrews, M. P. In Encyclopedia of Polymer Science and Engineering; John Wiley and Sons, Inc.; New York: 1987; vol. 9.

41. Andrews, M. P.; Ozin, G. A. J. Phys. Chem 1986, 90, 2922.
42. Klabunde, K. J. Paper Number 105, Am. Chem. Soc., National Meeting Anaheim, California, 1986.
43. Lin, S.-L.; Franklin, M. T.; Klabunde, K. J. Langmuir 1986, 2, 259.
44. Wada, N. J. de Physique, Colloque C2 1977, 38, 219.
45. Kimura, K. Bull. Chem. Soc. Japan 1984, 57, 1683; Kimura, K.; Bandow, S. Bull. Chem. Soc. Japan 1983, 56, 3578.
46. Meier, P. F.; Pennella, F.; Klabunde, K. J. J. Catal. 1986, 101, 545; Kanai, H.; Tan, B. J.; Klabunde, K. J. Langmuir 1986, 2, 760.
47. Schmidt, F. J. Mag. Mag. Mat. 1986, 54-57, 750.
48. Birringer, R.; Gleiter, H.; Klein, H.-P.; Marquardt, P. Phys. Lett. 1984, 102A, 365.
49. Togashi, S.; Fulcher, J. G.; Cho, B. R.; Hasegawa, M.; Gladysz, J. A. J. Org. Chem. 1980, 45, 3044.
50. Cook, N. D.; Timms, P. L. J. Chem. Soc. Dalton Trans. 1983, 239.
51. DeKock, C. W.; McAfee, L. V. Inorg. Chem. 1985, 24, 4293.
52. Davis, S. C.: Severson, S.; Klabunde, K. J. J. Am. Chem. Soc. 1981, 103, 3024.
53. Scott, B. A.; Plecenik, R. M.; Cargill, G. S.; McGuire, T. R.; Herd, S. R. Inorg. Chem. 1980, 19, 1252.
54. Kilner, M.; Mason, N.; Lambrick, D. B.; Hooker, P. D., Timms, P. L. J. Chem. Soc. Chem. Commun. 1987, 356.
55. Lianos, P.; Thomas, K. Chem. Phys. Lett. 1986, 125, 299 (and references cited therein); Boutonnet, M.; Kizling, J.; Touroude, R.; Maire, G.; Stenius, P. Appl. Catal. 1986, 20, 163.
56. O'Grady, K.; Bradbury, A. J. Mag. Mag. Mat. 1983, 39, 91, and references cited therein.
57. Francis, C. G. PhD. Thesis, University of Bristol, Bristol, England, 1978.
58. Nazar, L. F., Ph.D. Thesis, University of Toronto, Canada, 1986; Ozin, G. A.; Andrews, M. P. In Studies in Surface Science and Catalysis Gates, B. C.; Guczi, L.; Knozinger, H., Eds.; Elsevier: New York, 1986; ch. 7.
59. Andrews, M. P.; Ozin, G. A. U.S. Patent 4 452 913, 1984.
60. Klabunde, K. J. Ann. N.Y. Acad. Sci. 1977, 295, 83.
61. Klabunde, K. J.; Imizu, Y. J. Am. Chem. Soc. 1984, 106, 2721.
62. Radford, P. L.; Francis, C. G. J. Chem. Soc. Chem. Commun. 1983, 1520.
63. Akhmedov, V. L.; Anthony, M. T.; Green, M. L. H.; Young, D. L. J. Chem. Soc. Dalton Trans. 1975, 1412.
64. Mulhaupt, R.; Klabunde, U.; Ittel, S. J. Chem. Soc. Chem. Commun. 1985, 1745.
65. Walsh, D. J.; Higgins, J. S; Maconnachie, Eds. Polymer Blends and Mixtures Nato ASI Series E No. 89; Martinus Nijhoff Publishers: Boston, 1985; Noshay, A.; McGrath, J. E. Block Copolymers Academic Press: New York, 1977.
66. Heffner, S.; Andrews, M. P.; Galvin, M., to be sumitted.
67. J. Young, Ltd., 11 Colville Rd., Acton, London W3 8BS, England.

RECEIVED August 18, 1987

Chapter 7: Application 1

Heated Inlet System
for Cocondensation Metal Atom
Reactor

Thomas J. Groshens and Kenneth J. Klabunde

Department of Chemistry, Kansas State University, Manhattan, KS 66506

Many metal atom synthetic procedures utilizing a cocondensation apparatus (1) often require the use of compounds with low volatility and/or very high inlet rates. Without heating of the ligand inlet system such compounds condense on the inside of that system. If the compound is allowed to contact the hot metal vaporization source the resulting pyrolysis can seriously affect the purity and yield of the products. Therefore, it is necessary to have a means of maintaining sufficiently high temperature in the inlet system to volatilize the reactants. In our laboratory we find that wrapping the accessible portion of the inlet system with heating tape prevents condensation there, although another means of heating the showerhead portion which is inside the vacuum chamber is required.

The showerhead ligand inlet system is a 37cm piece of 14mm pyrex tubing closed at the end and having a series of one mm holes spaced about every 5mm around the bottom 7cm of the tube to provide an even distribution of the vapor on the reactor walls. Wrapping this tube with 180cm of 24 AWG Nichrome wire spaced one cm apart provides a heater with a resistance of about 10 ohms. In order to partially insulate and to hold the wire in place, a second 18mm Pyrex tube is placed around the upper portion above the area with the inlet holes (see Figure 1). Since the vaporization sources generally require 4-10 volts a convenient method of providing power to the heater is to connect the ends of the wire to the electrodes. This provides from 1.5 to 10 watts of heat to the showerhead system during the ligand-metal codeposition which works well for most compounds that have boiling points below about 150 degrees Celsius (at STP). With this setup the showerhead temperature depends on the metal vaporization source power input and cannot be independently controlled without altering the resistance of the system by changing the length of diameter of the heater wire. For instance the same 180 cm of 22 gauge wire has a resistance of 6.3 ohms, while 26 gauge wire has a resistance of 16 ohms. When inletting compounds with very high boiling points it is necessary to provide an independent voltage source to heat the showerhead. This is done with a separate pair

0097–6156/87/0357–0190$06.00/0
© 1987 American Chemical Society

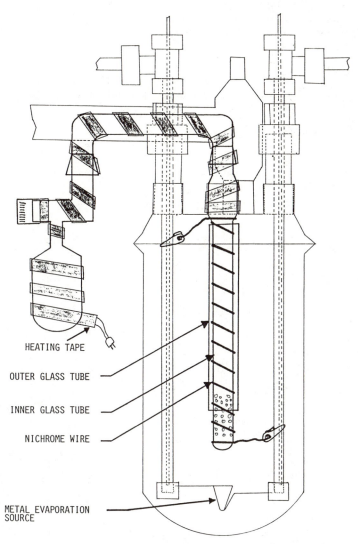

HEATING TAPE

OUTER GLASS TUBE

INNER GLASS TUBE

NICHROME WIRE

METAL EVAPORATION
SOURCE

Figure 1.   An outer glass tube provides partial insulation for the
inlet system, in addition to holding the heating wires in place.

of electrodes made of solid 1/8 inch Cu rods that are cemented into a ground glass joint with an epoxy resin and attached to the reactor at a convenient port. Using this method we have been able to use ligands such as 2,2'-bipyridine with a normal boiling point of 275 degrees Celsius at STP with no difficulty.

Literature Cited
1.   Klabunde, K. J.; Timms, P. L.; Skell, P. S.; Ittel, S.; Inorganic Syn., 1979, 19, 59-85, Shriver, D., editor, Wiley Interscience, New York, 1979.

RECEIVED September 1, 1987

Chapter 7: Application 2

Recovering Air-Sensitive Products from Metal Atom Reactors

Thomas J. Groshens and Kenneth J. Klabunde

Department of Chemistry, Kansas State University, Manhattan, KS 66506

After carrying out a metal atom reaction the product often needs to be syphoned out under airless conditions. In many cases the product is an insoluble solid slurried or suspended in a liquid organic solvent, for example supported metal catalysts or metal slurries. Such materials require a means of quick airless transfer that is not provided by normal Cannula techniques. We find that a 1.5M length of 0.25 inch O.D. thick wall (0.03 inch) Teflon tubing available from Aldrich provides a moderately flexible large diameter tube that can easily be attached to airless glassware via ACE threaded joints (#7 ACE-Thred Connector with compression O-ring). Slurries or solutions are easily transferred airlessly as follows: (1) the reactor is filled with N_2 and a top port opened under N_2 flush, (2) the Teflon tube, previously N_2 flushed and with a continuing N_2 flush, is connected to the airless glassware collection/filtration apparatus, and then inserted into the bottom of the reactor, (3) a slight vacuum is applied to the collection apparatus causing the slurry at the bottom of the reactor to be quickly sucked over (see Fig. 1). During this operation the reactor is continually under a N_2 flush.

RECEIVED September 24, 1987

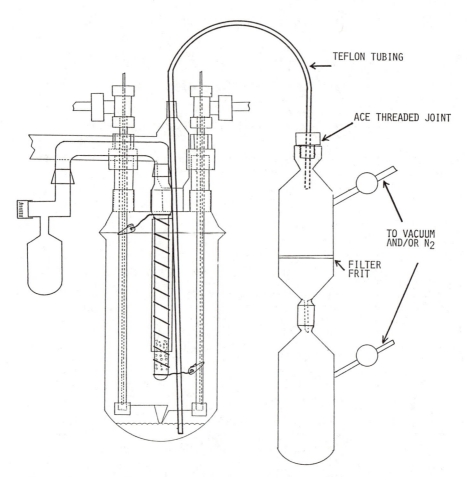

Figure 1.   The slurried reaction product is transferred aerobically
to the filter frit by applying a slight vacuum.

Chapter 7: Application 3

Versatile Sonochemical Reaction Vessels

Kenneth S. Suslick and Edward B. Flint

School of Chemical Sciences, University of Illinois at Urbana–Champaign, Urbana, IL 61801

Reaction vessels are described for the ultrasonic irradiation of air-sensitive solutions. These rigs are convenient for temperatures over the range of -50° to 150° and for pressures up to 10 atmospheres. One apparatus is described which can be used for sonoluminescence studies.

The use of ultrasound to promote and enhance chemical reactions has increased rapidly over the past few years [1-4]. The exclusion of air and water from these reactions is often critical, especially when organometallic compounds are being used [5-8]. The most commonly available source of ultrasound is the ultrasonic cleaning bath, which can be used with standard Schlenk techniques. While marginally effective with very reactive metals (e.g., Li, Mg, or Zn) [9-14], low intensity baths are insufficient to be of use for most sonochemical reactions. The most intense laboratory source is the ultrasonic immersion horn, which is commonly used by biochemists for cell disruption; commercial sources include Heat Systems-Ultrasonics, 1938 New Highway, Farmingdale, NY 11735, Branson Sonic Power, Eagle Road, Danbury CT 06810, and Sonics & Materials, Kenosia Ave., Danbury CT 06810. The introduction of samples in a gas tight manner and the control of temperature must be accomplished without damping the vibration of the immersion probe. For this reason, normal laboratory glassware cannot be used with this device. Our group has developed several reaction vessels for this type work, and we will describe two of them here.

A glass reaction vessel suitable for most sonochemical reactions is shown in Figure 1. The cell is made from 25 mm o.d. glass tubing that has been sealed at the bottom. The cell is held to the probe with a stainless steel collar that threads onto the probe itself. The threading is at the null point of the acoustic standing wave present in the probe, and thus does not dampen the motion of the probe tip. The o-rings at the collar-probe and cell-collar junctions provide a gas tight seal. Sidearms, made of 6 mm o.d. tubing, are capped with wired-on septa for the introduction or removal of gasses and samples under an inert atmosphere; alternatively, high-vacuum stopcocks or screw-cap teflon lined septa can be added for the most sensitive reactions. The volume of the glass cell can be easily varied from roughly 15 to 500 mL by replacing the cylindrical vessel with a round bottom flask. The entire cell is immersed in a cooling bath for reproducible temperature control. Using frozen slushes or a thermostated refrigerated bath, the range from -50° to

0097–6156/87/0357–0195$06.00/0

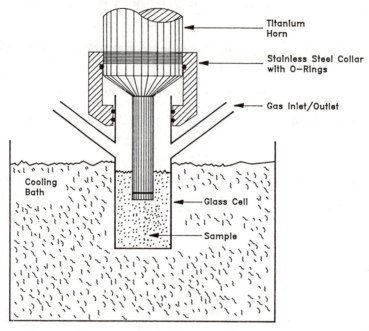

Figure 1. A glass reaction vessel suitable for most sonochemical reactions. The volume of the glass cell can be easily varied from 15 to 500 mL by replacing the cylindrical vessel with a round bottom flask.

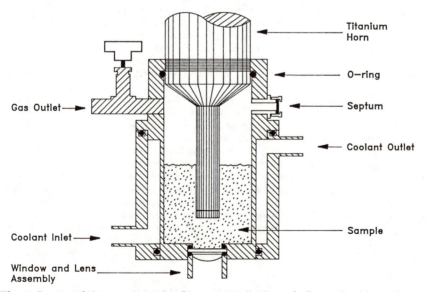

Figure 2. A stainless steel sonoluminescence cell. In a similar cell without the window and lens assembly and with a second valve in place of the septum fitting, one may operate above ambient pressure, with good pressure control up to 10 atmospheres.

150° is easily available. The temperature inside the reaction vessel must be monitored with a thermocouple <u>during</u> ultrasonic irradiation, since a significant rise in temperature will occur, reaching steady-state within a few minutes. As noted elsewhere [2], it is important that the total vapor pressure be kept low in order to maximize the efficacy of acoustic cavitation.

A stainless steel sonoluminescence cell is shown in Figure 2. This cell features a window at the bottom of the cell for the emitted light, inert gas ports, and a cooling jacket. The window is held in place by a holder threaded into the bottom of the cell. O-rings on either side of the window provide a leak proof seal and prevent the holder from damaging the window. The inside of the holder can accommodate a lens. The gas inlet is a septum through which a syringe needle may be used to sparge and sample the reaction solution. The gas outlet is a gas tight valve, which permits control of the internal cell pressure. In a similar cell with a flat metal bottom (<u>i.e.</u>, no window) and with a second valve in place of the septum fitting, one may operate at higher pressures, with good pressure control up to 10 atmospheres. The cooling jacket is threaded onto the outside of the cell for temperature control; alternatively, direct immersion into the cooling bath is possible.

Acknowledgments

We thank S. J. Doktycz, D. A. Hammerton, R. J. Johnson, P. F. Schubert, and H. H. Wang, who have all contributed to the development of these designs. Funding from the National Science Foundation is gratefully acknowledged. K.S.S. is a Sloan Foundation Research Fellow and the recipient of a Research Career Development Award of the National Institute of Health.

Literature Cited

1. Suslick, K. S. <u>Adv. Organometallic Chem.</u> **1986**, <u>25</u>, 73-119.
2. Suslick, K. S. <u>Modern Synthetic Methods</u> **1986**, <u>4</u>, 1-60.
3. Suslick, K. S. (Ed.) <u>Ultrasound: Its Chemical, Physical, and Biological Effects;</u> VCH Publishers: New York, 1987.
4. Suslick, K. S. in <u>High Energy Processes in Organometallic Chemistry;</u> Suslick, K. S., Ed.; ACS Symposium Series #333: Washington, D. C.; 1987, p. 191-208.
5. Suslick, K. S.; Goodale, J. W.; Wang, H. H.; and Schubert, P. F. <u>J. Am. Chem. Soc.</u> **1983**, <u>105</u>, 5781.
6. Suslick, K. S.; Schubert, P. F. <u>J. Am. Chem. Soc.</u> **1983**, <u>105</u>, 6042.
7. Suslick, K. S.; Johnson, R. E. <u>J. Amer. Chem. Soc.</u> **1984**, <u>106</u>, 6856.
8. Suslick, K. S.; Cline, Jr., R. E.; Hammerton, D. A. <u>J. Amer. Chem. Soc.</u> **1986**, <u>108</u>, 5641.
9. Luche, J. L. <u>Ultrasonics</u> **1987**, <u>25</u>, 40, and references therein.
10. Luche, J. L.; Damiano, J. C. <u>J. Amer. Chem. Soc.</u> **1980**, <u>102</u>, 7926.
11. Boudjouk, P.; Han, B. H.; Anderson, K. R. <u>J. Am. Chem. Soc.</u> **1982**, <u>104</u>, 4992.
12. Boudjouk, P. <u>J. Chem. Ed.</u> **1986**, <u>63</u>, 427, and references therein.
13. Boudjouk, P. in <u>High Energy Processes in Organometallic Chemistry;</u> Suslick, K. S., Ed.; ACS Symposium Series #333: Washington, D. C.; 1987, pp. 209-222.
14. Kitazume, T; Ishikawa, N. <u>Chem. Lett.</u> **1982**, 137; **1984**, 1453.

RECEIVED August 25, 1987

Chapter 7: Application 4

Convenient Pressure Reactors for Organometallic Reactivity Studies

Louis Messerle

Department of Chemistry, University of Iowa, Iowa City, IA 52242

Two glass reactors for 50-150 psi gas/solution reactions
are described. The first is a simple reactor with
safety pressure release, sample withdrawal port, and
quick connect hookup. The second incorporates a metal
pressure-equalizing addition funnel for reactant additions to the pressurized reaction mixture. Plans and
commercially--available parts lists for both designs
are provided.

Reactivity studies involving gaseous reagents and organotransition
metal species are usually done at either high pressures in commer-
cially--available stainless steel apparatus or at atmospheric pressure by dispersion of the gas in a solution of the organometallic
species in glass apparatus. Visible reaction changes, often important in judging reaction extent or completion, can be observed indi-
rectly in the former case only by sample withdrawal but directly in
the latter case. However, gas solubility at low pressures is often
a limiting factor in the latter method. There are few descriptions
of glass pressure reactors which cover the low pressure range of 50-
150 psi (1); most researchers utilize apparatus of local, and consequently unpublished, design.

In our research in low valent early transition metal organometallic chemistry, we have found the following two low pressure glass
reactors, both constructed of commercially available parts, to be
very useful in synthesis and reactivity studies. The simplest design
(Figure 1, right side) is a Fischer-Porter/Lab-Crest Scientific 3 oz.
aerosol pressure vessel (with associated metal couplings, but lacking
the cylindrical plastic shield for figure clarity) with a double-end
shut-off quick connect fitting, adjustable pressure relief valve, and
septum inlet/sample withdrawal port which is backed up with a plug
valve. This vessel can be loaded with an organometallic complex dissolved in a solvent by either glove box or Schlenk line techniques
(in the latter case by use of a female quick connect attached to a
rubber tubing hose connector) and then connected to a quick connect--
equipped gas manifold (containing a gauge, a connection to the gas
supply tank, and an evacuation/purge port). The adjustable pressure

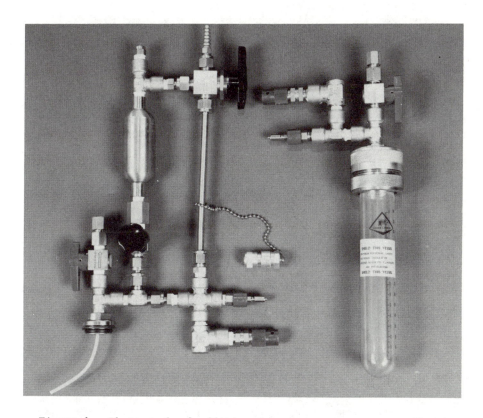

Figure 1. Photograph of addition funnel pressure reactor (left) and simple pressure reactor (right). The pressure vessel and associated couplings are omitted from the addition funnel design for clarity.

relief valve is an important safety feature which we set to 75 psi;
the unscratched bottles are presumably safe to 100 psi depending on
internal volume (Lab Crest does not certify these vessels for any
given pressure, so we operate routinely with safety shields around
the gas manifold). For added safety the user can purchase from Lab
Crest plastic cylindrical or stainless steel mesh shields for the
various volume (3, 6, or 12 fluid ounce) bottles. The plug valve
which backs up the septum inlet allows one to easily change the
septum during kinetic studies by partially opening the plug valve
while screwing on the septum retaining nut with a new septum. Construction of the apparatus from other materials (e.g., Monel) would
allow use with more corrosive gases.

Figure 1 also depicts a pressure reactor with integral pressure-
equalizing addition funnel (left portion of figure, with pressure
bottle and couplings removed for clarity). We have found the addition funnel feature to be extremely useful in low pressure addition
reactions (for example, in additions of alkali metal amalgams to
organotransition metal species in the presence of gaseous reactants).
Since common stainless steel alloys are inert to mercury, we have
constructed the entire apparatus from this material; the added cost
is more than compensated for by its increased versatility. As with
the simple reactor, this addition funnel--equipped reactor allows one
to withdraw samples during the course of a reaction. The three--way
ball valve allows one to isolate the sample cylinder area from the
reactor, fill it with a reagent solution using Schlenk procedures,
and then repressurize it prior to reagent addition. The design uses
a short piece of Teflon tubing, press--fitted into the port connector
located below the cylinder valve, to conduct the reagent solution
into the glass vessel. The sample cylinder is available in several
volumes, all of which can be accomodated easily by increasing the
length of stainless steel tubing in the pressure--equalizing branch.

Figure 2 shows a disassembled view of the simple reactor, while
the addition funnel reactor is shown in expanded view in Figure 3
(with the Lab Crest couplings omitted from the view for clarity).
Figure 3 shows the location of the Teflon tubing and the port connector, located below the shut off valve, into which the tubing is
press--fitted. Tables I and II list the required parts for both
reactor designs. The simple reactor and addition funnel reactor in
stainless steel (recommended) cost $341 and $547, respectively
(January 1987 prices), while the analogous prices in brass are $284
and $404. An appreciable portion of the cost is the 40% markup for
the purchase of unit quantities of the Lab Crest equipment from the
distributor, Aerosol Laboratory Equipment; Lab Crest Scientific only
sells directly in lots of one dozen. The pressure--equalizing addition funnel reactor can be constructed for $466 (stainless steel) if
the three--way ball valve is omitted. One valuable, though expensive
($42), feature is the substitution of a size 210 Kalrez O-ring for
the needle valve adapter O-ring supplied by Lab Crest. We have
noticed appreciable O-ring degradation and the need for frequent
replacement of common O-rings with solvents such as tetrahydrofuran,
while Kalrez is a fluoropolymer inert to all substances except molten
alkali metals.

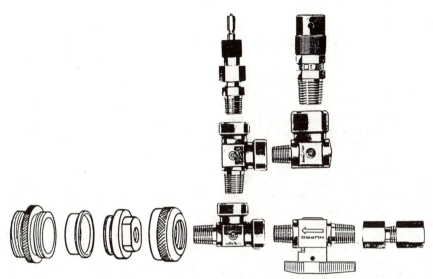

Figure 2. Expanded view of simple pressure reactor. The glass
pressure vessel is omitted for clarity. The septum retaining
nut is shown attached to the machined female Swagelok connector.
(Component illustrations reproduced with permission of Swagelok
Company and Fischer & Porter Company).

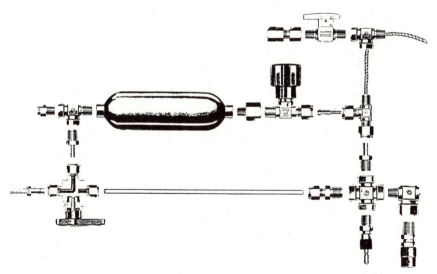

Figure 3. Expanded view of pressure--equalizing addition funnel
pressure reactor. The glass vessel and associated couplings are
omitted for clarity. The Teflon tubing is shown as the cross-
hatched tubing at lower right, while the stainless steel tubing
is depicted at left. (Component illustrations reproduced with
permission of Swagelok Company).

Table I. Parts List for Simple Pressure Reactor

Description	Part Number	Price
Street tee, Cajon, 1/8" NPT (2), @ $13.70	SS-2-ST	$ 27.40
Reducing street elbow, Cajon, 1/8" MPT to 1/4" FPT	SS-4-RSE-2	11.70
Plug valve, Nupro, 1/8" MPT, 4.4 mm orifice	SS-2P4T2	33.70
Female connector, Swagelok, 1/4" tube, 1/8" FPT (must have Swagelok end turned down approximately 1/16" on lathe in order to provide septum seating surface)	SS-400-7-2	6.05
Adjustable pressure relief valve, Nupro, 50-150 psi cracking pressure, 1/4" MPT	SS-4CPA2-50	21.40
Deflector cap for relief valve, Nupro, 1/4"	P-4CP4-P12-GR	2.20
Quick connect, Swagelok, double end shut off stem, 1/8" MPT	SS-QC4-D-2PM	15.70
GC injection port nut, Hewlett-Packard	18700-20030	18.00
GC septum (user specified)		
Needle valve adapter, Lab Crest, 1/8" FPT	110-957	28.00
Rubber washer, Lab Crest	110-973	1.99
Aerosol reaction vessel, Lab Crest, 3 oz. size without foot, with millimeter scale, coupling, and plastic shield (consists of: Aerosol coupling, 110-700 Aerosol coupling, 110-601 Polyethylene insert, 110-759 Reaction vessel, 110-452-0003 Cylindrical shield, 110-502-0003)	110-207-0003	175.00
Kalrez O-ring, size 210, for needle valve adapter, distributed by J. Crane (available in ≥ 10 quantity @ $36.30)	210-3765	42.20

Suppliers: Cajon/Nupro/Swagelok/Whitey: Swagelok Co., 29500 Solon
              Road, Solon, Ohio 44139; 216-248-4600
           Lab Crest Scientific/Fischer & Porter Co., County Line
              Road, Warminster, Pennsylvania 18974 (OEM for
              pressure reaction vessels and couplings, selling
              only in minimum quantities of one dozen; write for
              illustrated brochure describing vessel/shield
              options); 215-674-6610
           Aerosol Laboratory Equipment, RD-1, Route 10, Box 75,
              Walton, New York (distributor for unit quantities
              of Lab Crest equipment, at 40% markup; prices above
              are for this distributor); 607-865-7173
           Hewlett-Packard Co., c/o ASO Department, Route 41 & Starr
              Road, Avondale, Pennsylvania 19311; 800-227-9770
           J. Crane Houdaille, Inc., 6400 Oakton, Morton Grove,
              Illinois 60053; 312-967-2400

Table II. Parts List for Pressure Reactor with Integral
Pressure--Equalizing Addition Funnel

Description	Part Number	Price
Fractional tube adapter to male pipe, Cajon, 1/8" MPT, 1/4" tube (2), @ $2.70	SS-4-TA-1-2	$ 5.40
Three--way ball valve, Whitey, 1/4" Swagelok	SS-43XS4	65.50
Forged--body shut off valve, Whitey, 1/4" MPT, 1/4" Swagelok	SS-14DKM4-S4	38.20
Sample cylinder, Whitey, 40 mL, 1/8" FPT	304L-HDF2-40CC	43.80
Cross, Cajon, 1/8" FPT	SS-2-CS	18.60
Male run tee, Swagelok, 1/8" MPT, 1/4" tube	SS-400-3TMT	15.30
Reducing adapter, Cajon, 1/4" FPT, 1/8" MPT	SS-4-RA-2	4.80
Male connector, Swagelok, 1/8" MPT, 1/4" tube	SS-400-1-2	4.40
Tube hose connector, Cajon, 1/4" tube, 1/4" hose ID, tapered	SS-4-THC-4T	4.50
Port connector, Swagelok, 1/4" tube	SS-401-PC	3.65
Pipe plug, Cajon, 1/8" MPT	SS-2-P	1.80
Stainless steel tubing, 1/4" OD, 1 foot for 40 mL size sample cylinder		
Teflon tubing, 3 mm OD x 1 mm wall, 1 foot		
All parts listed in Table I		

Note: Teflon and Kalrez are trademarks of E.I. du Pont de Nemours and
Company; Nupro, Whitey, Cajon, and Swagelok are trademarks of
Swagelok Company. Component illustrations in Figures 2 and 3
are published by permission of the copyright owners.

Acknowledgments

We thank the U.S. Department of Energy, through the Pittsburgh Energy
Technology Center, for partial support of this work (DE-FG22-
85PC80513). Acknowledgment is made to the Donors of The Petroleum
Research Fund, administered by the American Chemical Society, for
partial support of this research (17193-G3).

Literature Cited

1. Bain, M. J.; Lavallee, D. K. J. Chem. Educ. 1976, 53, 221.

RECEIVED August 11, 1987

Chapter 8

Spectroscopic Characterization of Inorganic and Organometallic Complexes by Metal and High-Pressure NMR

D. Christopher Roe

Central Research and Development Department, Experimental Station, E. I. du Pont de Nemours & Co., Wilmington, DE 19898

Practical guidelines are offered for obtaining NMR spectra either of metals or of samples which are subjected to conditions of high pressure. Both techniques are becoming routinely available as a means for characterizing organometallic complexes. Applications of metal NMR are no longer limited to static characterization in terms of sample homogeneity or trends in chemical shift. Examples are drawn to emphasize structural characterization and dynamic studies. A novel sapphire NMR tube is described which permits routine high resolution operation up to 2,000 psi. The tubes can be used in any spectrometer, and have allowed the study of fundamental reactions of transition metal carbonyl complexes under conditions which would otherwise bring about sample decomposition.

The first part of this review introduces some of the practical considerations which may be required to obtain an NMR spectrum of a transition (or other) metal of interest. An underlying theme is that such studies are no longer limited to the determination of chemical shift or sample homogeneity, and literature examples (through mid-1986) have been selected with this more elaborate emphasis in mind. The second part of the review deals with a recent development in "high-pressure" NMR which facilitates the study of systems under pressures up to a few thousand psi. An advantage of the approach described is that it may be implemented relatively easily and requires no hardware modification to the spectrometer.

Metal NMR

The principal reason for the increasing importance of metal NMR as a means for sample characterization is simply the more general availability of commercial broad-band NMR spectrometers, with a typical range of RF components from 5 - 500 MHz. The range from low frequency metals, such as rhodium, on up to relatively high frequency nuclei such as phosphorus can typically be spanned using

0097–6156/87/0357–0204$06.00/0

just one or two probes, and the advent of wide-bore superconducting
magnets has permitted the advantage of large sample volumes (e.g.,
tubes 20 mm in diameter holding 10-12 mL).

Instrumental Considerations.   Of necessity for multinuclear
observation, one must be able to conveniently tune and match the
broad-band probe to the desired frequency. If software facilities
for accomplishing this are not directly available on the
spectrometer, the probe tuning curve can be displayed on an
oscilliscope using a sweep generator (e.g., Wavetek Model 10621) and
a tuning bridge (e.g., Wiltron Model 62BF50). Externally adjustable
capacitors in the probe are then adjusted to tune the probe response
to the desired frequency, and to match the response to a 50-ohm load
(1). The high-Q probes typically employed in high resolution NMR
are sufficiently sensitive that a well-tuned probe at room
temperature may become detuned by more than 1 MHz by changing the
temperature to -80°C. Similar effects may be noted for samples
having substantially different dielectric properties, and failure to
account for these tuning differences may lead to longer pulse
lengths and markedly degraded S/N.

Once an operating frequency has been chosen, the search for
resonance is facilitated by using the widest available sweep width
(typically around 100 KHz, depending on digitizer speed) and
quadrature detection. However, 90° pulse lengths typical for high
resolution spectrometers are not capable of providing uniform
excitation across such spectral widths (2) and, in fact, regions of
null excitation occur at offset frequencies related to 1/(pulse
length). The chances of observing a signal which occurs in the
sweep window can therefore be improved by using smaller pulse widths
(e.g.,10 μs) and relatively rapid recycle times. Once the chemical
shift region of interest has been defined, it may be desirable to
operate the spectrometer frequency close to resonance and, for
certain more sophisticated experiments, to determine accurate 90°
pulse lengths. The latter task is greatly aided if a
readily-detected set-up sample can be found (preferably with
dielectric properties similar to that of the sample of interest).
In general, 90° pulse lengths less than 50 μs are highly desirable.

Superconducting magnets conventionally have vertically-mounted
(Helmholtz coil) probes which permit ready sample access, but which
also have some undesirable features. The alternative is a
sideways-mounted solenoidal probe, which typically gives
approximately a factor of 2 improvement in S/N and markedly better
H_1 pulse characteristics. For high-loss samples (e.g., concentrated
salt solutions in water), H_1 homogeneity may be so poor in a
conventional probe that it is impossible to achieve a 180° pulse,
and in such cases a solenoidal probe would be required for T_1 or
2-dimensional experiments. Thus, for special cases involving low
solubility or sensitivity, or unacceptably long pulse lengths, a
sideways-mounted probe may offer sufficient advantage to merit the
associated expense, the inconvenience of sample loading, and the
difficulty of shimming an unconventional probe orientation.

One of the principal goals of probe and amplifier design is to
produce short pulse lengths both for good power distribution
characteristics, and also for improved performance of the more
demanding pulse sequences which will find increasing use in

multinuclear NMR (3). Since these pulse sequences can involve 90°
and 180° pulses on both the observe and the decoupler channels, a
reasonably sophisticated pulse programmer is required for their
implementation. Elaborate pulse sequences are also required in
efforts to minimize the effects of acoustic ringing, which can be a
sufficiently serious problem below about 20 MHz that it can
overwhelm the signal of interest. The sources of acoustic ringing
have been described (2,4), and, of the solutions proposed, a
particularly effective one given is given by Ellis (Ellis, P. D.,
discussion at the Nato Meeting Advanced Study Institute on
Multinuclear NMR Spectroscopy, Stirling, 1981; cf (5)); a brief
discussion and further apparent improvements are given by Eckert and
Yesinowski (6). A number of generally useful hints and techniques
are discussed in (7).

Nuclear Properties Affecting Detectability. Among the difficulties
that can be experienced in trying to observe a given metal is the
extremely large range of chemical shifts which may occur; a notable
example is Pt, whose chemical shift range spans about 13,000 ppm.
While the literature may provide guidelines in locating a metal
chemical shift for a particular oxidation state (8-11), it may
nevertheless be required to search a number of spectral windows
because of pulse power and sweep width limitations. It should also
be noted that reliable chemical shift comparisons require
temperature regulation since, in some cases, shifts can vary by as
much as 1 ppm per degree.

While spin-1/2, or dipolar, metals are often associated with
narrow resonance lines, this useful feature is sometimes obtained
only at the expense of long T_1 relaxation times. Some improvement
in detectability may result from broad-band ^1H decoupling, although
the observed NOE depends on the extent to which dipolar interactions
contribute to the relaxation mechanism, and is unlikely to approach
its maximum possible value $\gamma_H/2\gamma_M$. A small NOE can actually lead
to a dereased or null signal for negative-γ metals such as Sn, Cd,
Rh, etc. In cases which involve resolved coupling to protons, an
attractive alternative involves polarization transfer experiments
such as DEPT (3,12). While this approach requires calibration of
decoupler pulse lengths for a given output level (2,13), the
advantages include enhancements greater than those provided by NOE
and repetition rates related to the proton T_1 (usually shorter than
the metal T_1). The enhancement is proportional to γ_H/γ_M, and is
independent of the dipolar contribution to the relaxation mechanisms
of the metal.

In practice, polarization techniques such as DEPT are not
generally suitable for quadrupolar nuclei. Quadrupolar relaxation
tends to be very efficient in the nonsymmetric environments likely
to be encountered in cases of synthetic interest, and this
relaxation effectively decouples any nearby protons. The typical
result of this relaxation is a broad unresolved line whose S/N ratio
can be built up only by rapid pulse repetition rates. The factors
involved in quadrupolar and other types of relaxation are
informatively discussed by Kidd (14).

Examples Involving Dipolar Metals. The dipolar metals are grouped
in Table I along with a rough indication of their ease of detection.

Typical work with the "difficult" group of nuclei would require large sample volumes and high solubility. A notable absentee from the Table is Yb(II); although it is moderately receptive (in principle), and resonates at a convenient frequency, I am unaware of any successful observation of this nucleus in solution (despite a number of known attempts).

There is an enormous literature concerning the dipolar metals, and earlier reviews (8,10,15) constitute an important source of reference information. Recent contributions include trends in Pt (16) and Sn (17) chemical shifts, and in Sn-X coupling constants

Table I. Classification of Dipolar Metals[a]

Easy	^{119}Sn, ^{195}Pt, ^{205}Tl
Moderately difficult	^{89}Y, ^{77}Se, ^{125}Te, ^{113}Cd, ^{199}Hg, ^{207}Pb
Hard	^{183}W, ^{103}Rh, ^{109}Ag, ^{57}Fe, ^{187}Os

[a] The ordering is thought to reflect a combination of ease of detection and degree of utility. ^{57}Fe is extremely difficult to observe without enrichment, and ^{187}Os is even less receptive than its quadrupolar spin isotope.

(17-21). Studies of Pt-Sn complexes frequently combine NMR observation of both nuclei (22,23), and use of both these probes has been of evident value in homogeneous catalysis (24,25). Of perhaps greater interest is the fact that tin NMR has been used directly to study ligand dynamics. Two-dimensional exchange spectroscopy (NOESY) was used to elucidate in a qualitative manner the mechanism of isomerization of the ditin complex $CH_2[(C_6H_5)Sn(SCH_2CH_2)_2NCH_3]_2$ (26). A quantitative approach to this sort of isomerization is therefore possible, either by means of phase-sensitive 2D NOESY (27) or by a complete set of 1D magnetization transfer experiments (28). Tin magnetization transfer has in fact been elegantly applied to a quantitative evaluation of cis-trans isomerization and Me_2S ligand exchange in $SnCl_4 \cdot 2Me_2S$ (29).

Variable temperature NMR lineshapes of metal nuclei can also provide information relating to the mechanism of dynamic processes. Fluxional behavior in complexes I and II can easily be followed

$$\begin{array}{c} PPh_3 \\ | \\ (PPh_3)-Pt-H \\ | \\ CpM(CO)_3 \end{array} \qquad \begin{array}{l} I \quad M = Mo \\ II \quad M = W \end{array}$$

by ^1H or ^{31}P NMR, but the observed lineshapes do not immediately suggest what sort of process should be used to model the exchange. Such ambiguities are removed by the simplicity associated with the variable temperature Pt NMR spectra shown in Figure 1. The fact that the outside lines of the spectrum remain sharp throughout the VT range indicates that the process involves a mutual exchange of the phosphines: the cis-trans isomerization must proceed in such a way that both the hydride and phosphine ligands remain attached to the metal. While the Pt lineshapes also provide a reasonable estimate of the exchange rates, more accurate rate information may be derived from the corresponding ^{31}P spectra in which the intrinsic linewidths are narrower.

A different sort of dynamic effect can be encountered when observing a spin-1/2 metal which is scalar coupled to a quadrupolar nucleus. In general, the lineshape of such a spin-1/2 nucleus I depends (i) on the T_1 of the quadrupolar nucleus S and (ii) on the magnitude of the scalar coupling J_{IS}, such that narrow lines observed at low temperature become increasingly broad as the temperature is increased (in contrast to the normally observed behavior). In the high temperature region, the long quadrupolar T_1 corresponds to a slow relaxation rate, and if this rate is slow relative to J, the spin-1/2 nucleus I sees 2S+1 equally populated sites; under these conditions, the spectrum of I consists of 2S+1 equally intense lines. The vanadium-coupled tungsten spectra illustrated in Figure 2 represent an intermediate case in which the quadrupolar vanadium nucleus is only partially "decoupled" from the tungsten at the highest accessible temperature (30). In addition to the variation of the measured quadrupolar T_1, the simulations in Figure 2 include an additional small exchange rate contribution required to provide the optimal fit to the data. Chemical exchange can affect these lineshapes in a characteristic way, but in practice the exchange effects tend to be rather subtle in comparison with the quadrupolar effects.

In systems containing a number of metal atoms, questions concerning connectivity can be answered in principle if J coupling between the metals can be observed. The structural characterization of heteropolyanions, for example, has been enormously aided by modern NMR techniques which rely for their success on the presence of scalar coupling. For example, the similarity of W-W coupling constants in $Li_5[PV_2W_{10}O_{40}]$ leads to ambiguity in the normal 1-D ^{183}W spectrum, whereas the 2D-INADEQUATE results shown in Figure 3 permit a unique assignment of the possible positional isomers (30).

Additional references of interest may be noted for ^{89}Y (31), ^{77}Se (32,33), ^{125}Te (34), ^{113}Cd (35-38), ^{103}Rh (39), ^{199}Hg (40) and ^{207}Pb (41).

Examples Involving Quadrupolar Metals. The quadrupolar metals are grouped in Table II, again with an indication of their ease of detection. The classification is considerably more arbitrary than for the dipolar metals since the ease of detection will depend markedly on the degree of symmetry around the metal and the magnitude of the quadrupole coupling constant. Where a choice exists, the most likely preferred isotope is indicated.

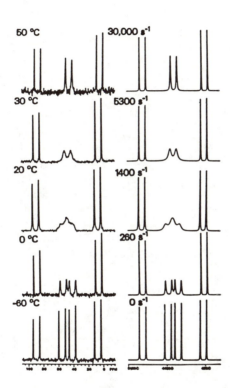

Figure 1.   Observed and calculated Pt NMR spectra for I at indicated temperatures (A. H. Janowicz and D. C. Roe, unpublished data).

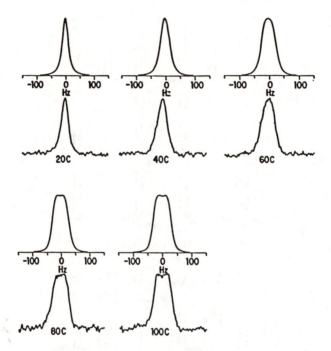

Figure 2.   Observed and calculated lineshape of the -81.1 ppm [183]W NMR resonance of $Li_5[SiVW_{11}O_{40}]$ as a function of temperature.   Simulations use the values $^2J_{W-O-V} = 11.5$ Hz and $t_{ex} = 40$ s^{-1}.

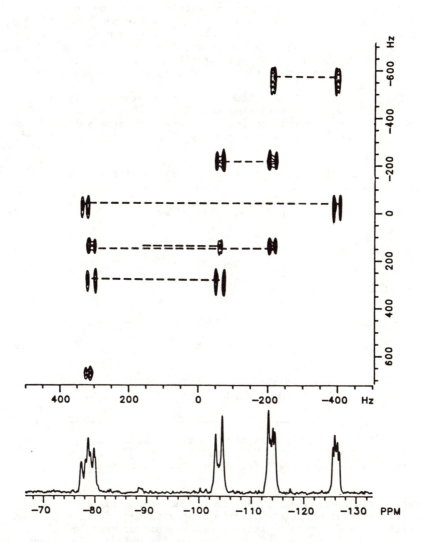

Figure 3.   2D-INADEQUATE ^{183}W NMR spectrum (^{51}V-decoupled) of α-1,2-Li$_5$[PV$_2$W$_{10}$O$_{40}$], 0.4M in D$_2$O at 30°C.

As is the case for the dipolar metals, the number of applications involving quadrupolar metals is far too great to cover in a short review such as this. Dechter's coverage (10) of the literature from 1978 through September, 1983 is particularly timely in updating earlier reviews, and selected references which are thought to be currently representative will simply be mentioned here.

Trends in the ^{59}Co chemical shifts of a wide variety of organocobalt complexes have been usefully interpreted in terms of structure-activity relationships in homogeneous catalysis (42) and in terms of the influence of various pi-ligands on structure and

Table II. Classification of Quadrupolar Metals[a]

Relatively easy	^{7}Li, ^{23}Na, ^{27}Al, ^{45}Sc, ^{51}V, ^{59}Co, ^{93}Nb
Moderately difficult	^{9}Be, ^{47,49}Ti, ^{53}Cr, ^{71}Ga, ^{87}Rb, ^{91}Zr, ^{95}Mo, ^{99}Ru, ^{133}Cs
Very difficult	^{25}Mg, ^{39}K, ^{43}Ca, ^{55}Mn, ^{61}Ni, ^{63}Cu, ^{67}Zn, ^{73}Ge, ^{75}As, ^{87}Sr, ^{99}Tc, ^{121}Sb, ^{137}Ba, ^{139}La
Useless[b]	^{105}Pd, ^{115}In, ^{175}Lu, ^{177}Hf, ^{181}Ta, ^{187}Re, ^{189}Os, ^{193}Ir, ^{197}Au, ^{209}Bi, ^{235}U

---

[a] The ordering can only be approximate. [b] The nuclear properties of this group make it extremely unlikely that NMR spectra will provide useful information for organometallic complexes.

chemical stability (43). ^{27}Al NMR has naturally figured in the characterization of organoaluminum compounds and, for example, has permitted the distinction between octahedral and tetrahedral sites in a series of alkoxide and siloxide complexes (44). An apparently interesting dynamic equilibrium between four- and five-coordination in diorgano[(2-pyridyl)methoxy]aluminum complexes (45) has been discounted in that the observations leading to this conclusion stem in part from a spurious signal found to be present when the sample tube contained only solvent (46). This example focuses attention on a general caveat that must be heeded when observing such broad resonances from quadrupolar metal nuclei: the absence (or presence) of background signals arising from the probe or sample tube alone must be confirmed prior to signal identification and interpretation.

In favorable cases, aspects normally associated with spin-1/2 systems can feature in the study of quadrupolar metals. For example, variable temperature ^{45}Sc NMR spectra provided evidence for a monomer-dimer equilibrium for Cp_2Sc in toluene (47). In addition, homonuclear J-correlation (COSY) experiments have been successfully applied to determine vanadium connectivites via $^2J_{V-0-V}$ in heteropolyanions, even when the breadth of the resonances obscures the spin-spin coupling (30). In the case of $Na_{9-x}H_x[PV_{14}O_{42}]$, the off-diagonal peaks correlating the resonances at -577 and -593 ppm (Figure 4) indicate that these latter peaks derive from the main Keggin framework, while the peak at -526 ppm shows no such correlation and is assigned to vanadium in the capping group. Such detection of connectivity patterns through J coupling in otherwise broad, featureless spectra may have implications for structural studies of other quadrupolar nuclei where it is possible to trade a problem of inherent resolution for a problem of sensitivity.

Additional references are noted for ^{99}Ru (48), ^{95}Mo (49-51), ^{87}Rb (52), ^{55}Mn (53), ^{99}Tc (54), ^{139}La (55), ^{63}Cu (56) and ^{25}Mg, ^{43}Ca, ^{39}K, and ^{67}Zn (57). In the latter case, use is made of isotopically enriched metals to overcome the low natural abundance of the NMR preferred isotopes for Mg, Ca and Zn, and in this way to study the interaction of these metals with the regulatory protein, calmodulin.

It is hoped that the above examples provide an indication of the role metal NMR can play in characterizing certain organometallic and inorganic complexes. In favorable cases, it may be seen that metal NMR provides the most direct insight to the structural details of multinuclear systems, and to the dynamics and mechanism of exchange processes occurring around the metal center.

High Pressure NMR

The extent to which pressure has been neglected as an experimental variable is due, in large part, to the lack of ready availability of appropriate pressure cells. This portion of the review describes a means by which high resolution NMR spectra can be obtained for samples subjected to pressures up to 2000 psi with what is considered to be a reasonable margin of safety. The key to this accomplishment is to take advantage of the tensile strength and other properties of sapphire tubing. While the emphasis is on these new sapphire NMR tubes, the relationship to other high pressure NMR techniques will be described first.

Conventional glass NMR tubes can often be used up to 10 atmospheres (ca. 150 psi), but they become increasingly unreliable at higher pressures and low temperatures. Repeated use which leads to the development of micro scratches on the surface of the glass can also lead to failure under conditions which had previously been successfully achieved. Nevertheless, this approach may be very convenient for work with gases such as H_2, CO and O_2, and heavy-walled glass tubes fitted with a symmetrical Teflon valve are commercially available (e.g., from Wilmad or other sources). The valve can be mated to a standard Swagelok fitting on a 1/16" line from a cylinder, and these devices provide perhaps the most convenient entry to a useful, albeit limited, pressure region.

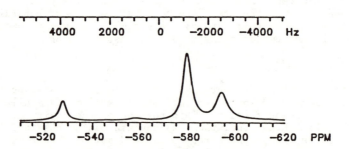

Figure 4. 2-D ^{51}V-^{51}V COSY spectrum of $Na_{9-x}H_x[PV_{14}O_{42}]$ at 60°C.

Very high pressures (e.g., in excess of 2000 atm) can be achieved by means of high pressure probes, but these are relatively demanding to construct (58) and tend to require a full-time commitment. The rewards that can be achieved by such efforts are exemplified by the elegant studies of the dynamic structure of liquids (59) or the conformational isomerization of cyclohexane (60), and by the estimation of volumes of activation for a number of chemical processes (58,61).

Intermediate pressures of 10 - 100 atm are of interest in that they are associated with the chemistry of reactant gases in solution, where such pressures permit a corresponding increase in the accessible concentrations of the gases compared to those available in the glass tubes described above. Areas where this sort of capability might be useful include kinetic studies which require the systematic variation of gas concentration, and studies involving reactive intermediates which decompose by dissociative loss of ligands such as CO. Another sort of application might involve supercritical fluids which can be maintained at modest pressures.

Construction and Operation of the Sapphire NMR Tube. Sapphire was chosen as the material for tube construction because its excellent tensile strength characteristics can be uniformly retained in a tube grown intact as a single crystal. Tubes 5 mm in o.d., with 0.8 mm wall and sealed at one end, were chosen for our prototype and were purchased from Saphikon, Inc. (51 Powers Street, Milford, NH 03055); at the time of our original purchase, the cost of the tubes was approximately $300 each.

The open end of the tube is sealed by means of a nonmagnetic titanium alloy valve (Figure 5). The numbered details are 1, valve body; 2, valve stem drive handle; 3, stem drive and packing gland; 4, nonrotating stem; 5, packing assembly; 6, gas inlet port for 1/16 in. tubing; 7, assembly screw (total of 4); 8, Viton O-ring seal; 9, tube mounting flange; 10, epoxy sealant; 11, spinner turbine; and 12, sapphire tube. The base plate or flange for the valve is epoxied to the tube (62) and the upper valve assembly is attached by means of four threaded bolts. The original valve design has been reduced to 25 mm o.d. in order to accommodate the dimensions of conventional bore superconducting magnets, and the weight of the valve and tube assembly is approximately 74 g. Samples can be syringed into the tube through the opening in the flange, and the valve bolted in place; for air-sensitive compounds, these operations are conveniently carried out inside a drybox.

The tube is mounted on the spinner and housed inside a safety shield consisting of a polymethylmethacrylate cylinder, baseplate and cap; the cylinder is long enough to accommodate the tube, spinner and lower portion of the valve. With the spinner supported by a pair of plastic bolts across the cylinder, the upper valve assembly is accessible for pressurization from the appropriate gas cylinder. The gas is introduced through 1/16" stainless steel tubing fitted with a standard Valco 1/16" ferrule and male nut which screws into the side of the valve. Rotation of the valve stem handle effects the opening and closing of the valve. Once the sample is sealed under pressure, it may be detached from the line and gently rocked on its side in order to mix the gas with the solution.

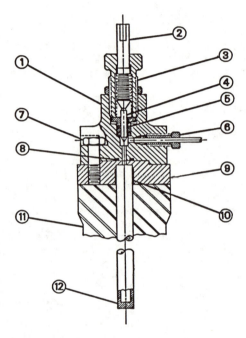

Figure 5. Schematic drawing of Ti-alloy valve and sapphire tube assembly.

The pressurized sample can be introduced into the probe without direct exposure to the tube. The baseplate and cap of the safety shield are removed, and the remaining assembly is made to rest on top of the magnet above the probe stack. The tube is restrained by means of a removable loop of string or lasso around the stem drive (part 3 in Figure 5) while the plastic bolts supporting the spinner are removed. The tube can then be lowered to the top of the probe stack where it can be supported by the eject air system and lowered into the probe.

Performance of the Sapphire NMR Tube. The tube and valve are sufficiently light weight and symmetrical that they spin and achieve a typical resolution of about 1 Hz. In hydrostatic pressure testing, the burst pressure of one tube was found to be approximately 14,500 psi, and it is therefore believed that operations up to 2000 psi may be carried out with a reasonable margin of safety. Obviously however, all due caution must be exercised and operator exposure to a pressurized tube must be avoided. It is as yet unclear whether extended use will eventually affect the performance of these tubes. In our hands, the tubes have been operated without incident, over a period of two years, from −140 to +150°C at nominal pressures up to 1200 psi.

Examples of High Pressure Studies. The sapphire tubes have been used to study CO exchange rates with $Co_2(CO)_8$ at pressures up to 1100 psi (63). ^{13}C NMR spectra obtained between 40 and 80°C reveal that free CO in solution is exchanging slowly with the CO's coordinated to cobalt. Since lineshape information at these and higher temperatures is complicated by broadening caused by the quadrupolar cobalt nucleus, magnetization transfer techniques were used to measure the exchange rates as a function of temperature and pressure. Carbon monoxide 99% ^{13}C-enriched was used in order to overcome the low S/N associated with the potentially long ^{13}C T_1's of the coordinated CO. Consistent with a small value for the equilibrium dissociation constant (Equation 1), the observed exchange rates were found to be independent of pressure and

$$Co_2(CO)_8 \rightleftharpoons Co_2(CO)_7 + CO \tag{1}$$

therefore to correspond directly to rates of dissociation. These rate constants lead to the activation parameters $\Delta H^{\ddagger} = 25.5$ (1.5) kcal/mol and $\Delta S^{\ddagger} = 22$ (4) e.u.

In a series of similar experiments, it was found that $CH_3CO-Co(CO)_4$ could be sufficiently stabilized by CO pressures between 100 and 1100 psi that it could be studied over a period of days at temperatures up to 80 or 90° (see Figure 6). Exchange of the terminal CO's with free CO takes place a little more rapidly than for $Co_2(CO)_8$, and for data obtained between 40 and 80° the observed rate constants for dissociation correspond to the activation parameters $\Delta H^{\ddagger} = 22.0$ (0.2) kcal/mol and $\Delta S^{\ddagger} = 8$ (0.5) e.u. Comparison of these results with those found for $Co_2(CO)_8$ suggests that loss of CO from $CH_3CO-Co(CO)_4$ is accompanied by an increase in structure of the transition state; presumably, this increase in structure corresponds to a stabilizing coordination of the acyl oxygen in η^2 fashion. In addition to the dissociation

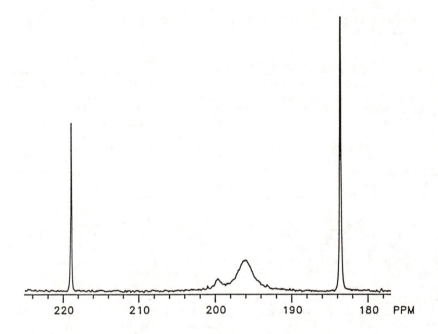

Figure 6. ^{13}C NMR spectrum of $CH_3CO-Co(CO)_4$ (218.9 and 196.2 ppm) obtained at 80°C in methylcyclohexane-d$_{14}$, 1000 psig. Free CO is at 183.6 ppm, and the small peak around 200 ppm is due to $Co_2(CO)_8$.

process, a slower scrambling of the acyl and terminal carbonyls can also be observed. As expected for deinsertion from the coordinatively unsaturated intermediate $CH_3CO-Co(CO)_3$, the acyl exchange rate is found to be markedly inhibited by CO pressure.

The same sapphire tubes can be used to characterize transition metal radical species by ESR. For example, it had previously been determined under ambient pressures and very low temperature that di- and tri-nuclear iron carbonyl radicals were formed during the photolysis of $Fe(CO)_5$ (Equation 2). The extremely reactive monomeric intermediate $HFe(CO)_4\cdot$ could be characterized ([64]) at

$$Fe(CO)_5 \xrightarrow[h\nu]{H_2} H\overset{\cdot}{F}e_2(CO)_8 + H\overset{\cdot}{F}e_3(CO)_{11} \tag{2}$$

$-110°C$ under 900 psi of 1:1 H_2/CO (Equations 3 and 4). Under these

$$Fe(CO)_5 \xrightarrow[h\nu]{H_2/CO} H_2Fe(CO)_4 \tag{3}$$

$$H_2Fe(CO)_4 \xrightarrow[h\nu]{-H\cdot} H\overset{\cdot}{F}e(CO)_4 \tag{4}$$

conditions the half-life of this 17-electron species is approximately 12 seconds. The diamagnetic precursor $H_2Fe(CO)_4$ was characterized by ^1H NMR at $-80°C$ using the same sample tube but with a more concentrated solution. Variation of the gas pressure and composition allowed the identification of additional hydride radicals.

It is intriguing to speculate that sapphire NMR tubes may also prove useful in spectroscopic studies involving supercritical fluids. It has been shown ([65]) that the very low viscosities of supercritical fluids provide an effective means for achieving line narrowing of quadrupolar nuclei such as metals. While it is premature to comment on successful applications in this area, the utility of these sapphire NMR tubes has been established for studying the reactions of organometallics with certain gases. Given the relative convenience of working with sapphire tubes, it may be anticipated that they will find use in a variety of applications which require operation at elevated pressure.

Literature Cited

1. Daugaard, P.; Jakobsen, H. J.; Garber, A. R.; Ellis, P. D. J. Magn. Reson. 1981, 44, 224.
2. Brevard, C. In NMR of Newly Accessible Nuclei; Laszlo, P., Ed.; Academic: New York, 1983, Vol. 1, Chapter 1.
3. Bolton, P. H. In NMR of Newly Accessible Nuclei; Laszlo, P., Ed.; Academic: New York, 1983, Vol. 1, Chapter 2.
4. Fukushima, E.; Roeder, S. B. W. J. Magn. Reson. 1979, 33, 199.
5. Gerothanassis, I. P.; Lauterwein, J. J. Magn. Reson. 1986, 66, 32.

6. Eckert, H.; Yesinowski, J. P. J. Am. Chem. Soc. 1986, 108, 2140.
7. Fukushima, E.; Roeder, S. B. W. Experimental Pulse NMR: A Nuts and Bolts Approach; Addison-Wesley: Reading, Massachusetts, 1981, Chapter 5, 6.
8. Kidd, R. G.; Goodfellow, R. J. In NMR and the Periodic Table; Harris, R. K.; Mann, B. E., Eds.; Academic: New York, 1978, Chapter 8.
9. Brevard, C.; Granger, P. Handbook of High Resolution Multinuclear NMR; John Wiley: New York, 1981.
10. Dechter, J. J. In Progress in Inorganic Chemistry; Lippard, S. J., Ed.; John Wiley: New York, 1985, Vol. 33, p. 393.
11. Webb, G. A. In NMR of Newly Accessible Nuclei; Laszlo, P., Ed.; Academic: New York, 1983, Vol. 1, Chapter 4.
12. Morris, G. A. In Topics in Carbon-13 NMR Spectroscopy; Levy, G. C. Ed.; John Wiley: New York, 1984, Vol. 4, Chapter 7.
13. Bax, A. J. Magn. Reson. 1983, 52, 76.
14. Kidd, R. G. In NMR of Newly Accessible Nuclei; Laszlo, P., Ed.; Academic: New York, 1983, Vol. 1, Chapter 5.
15. Benn, R.; Rufinska, A. Angew. Chem. Int. Ed. Engl. 1986, 25, 861.
16. Benn, R.; Reinhardt, R.-D.; Rufinska, A. J. Organomet. Chem. 1985, 282, 291.
17. Wrackmeyer, B. In Annual Reports on NMR Spectroscopy; Webb, G. A., Ed.; Academic: New York, 1985, Vol. 16, p. 73.
18. Howard, W. F. Jr.; Crecely, R. W.; Nelson, W. H. Inorg. Chem. 1985, 24, 2204.
19. Yano, T.; Nakashima, K.; Otera, J.; Okawara, R. Organometallics 1985, 4, 1501.
20. Lockhart, T. P.; Manders, W. F.; Brinckman, F. E. J. Organomet. Chem. 1985, 286, 153.
21. Molloy, K. C.; Quill, K.; Blunden, S. J.; Hill, R. J. Chem. Soc. Dalton Trans. 1986, 875.
22. Albinati, A.; Moriyama, H.; Ruegger, H.; Pregosin, P. S. Inorg. Chem. 1985, 24, 4430.
23. Albinati, A.; Von Gunten, U.; Pregosin, P. S.; Ruegg, H. J. J. Organomet. Chem. 1985, 295, 239.
24. Anderson, G. K.; Clark, H. C.; Davies, J. A. Inorg. Chem. 1983, 22, 427.
25. Anderson, G. K.; Clark, H. C.; Davies, J. A. Inorg. Chem. 1983, 22, 434.
26. Wynants, C.; Van Binst, G.; Mugge, C.; Jurkschat, K.; Tzschach, A.; Pepermans, H.; Glelen, M.; Willem, R. Organometallics 1985, 4, 1906.
27. Johnston, E. R.; Dellwo, M. J.; Hendrix, J. J. Magn. Reson. 1986, 66, 399.
28. Gesmar, H.; Led, J. J. J. Magn. Reson. 1986, 68, 95.
29. Knight, C. T. G.; Merbach, A. E. Inorg. Chem. 1985, 24, 576.
30. Domaille, P. J. J. Am. Chem. Soc. 1984, 106, 7677.
31. Evans, W. J.; Meadows, J. H.; Kostka, A. G.; Closs, G. L. Organometallics 1985, 4, 324.
32. Hope, E. G.; Levason, W.; Murray, S. G.; Marshall, G. L. J. Chem. Soc. Dalton Trans. 1985, 2185.

33. Allen, D. W.; Nowell, I. W.; Taylor, B. F. J. Chem. Soc. Dalton Trans. 1985, 2505.
34. Jones, C. H. W.; Sharma, R. D. Organometallics 1986, 5, 805.
35. Frey, M. H.; Wagner, G.; Vasak, M.; Sorensen, O. W.; Neuhaus, D.; Worgotter, E.; Kagi, J. H. R.; Ernst, R. R.; Wuthrich, K. J. Am. Chem. Soc. 1985, 107, 6847.
36. Keller, A. D.; Drakenberg, T.; Briggs, R. W.; Armitage, I. M. Inorg. Chem. 1985, 24, 1170.
37. Dean, P. A. W.; Vittal, J. J. Inorg. Chem. 1986, 25, 514.
38. Summers, M. F.; van Rijn, J.; Reedijk, J.; Marzilli, L. G. J. Am. Chem. Soc. 1986, 108, 4254.
39. Mann, B. E.; Meanwell, N. J.; Spencer, C. M.; Taylor, B. F.; Maitlis, P. M. J. Chem. Soc. Dalton Trans. 1985, 1555.
40. Bach, R. D.; Vardhan, H. B.; Rahman, A. F. M. M.; Oliver, J. P. Organometallics 1985, 4, 846.
41. Wrackmeyer, B. J. Magn. Reson. 1985, 61, 536.
42. Bonnemann, H. Angew. Chem. Int. Ed. Engl. 1985, 24, 248.
43. Benn, R.; Cibura, K.; Hofmann, P.; Jonas, K.; Rufinska, A. Organometallics 1985, 4, 2214.
44. Wengrovius, J. H.; Garbauskas, M. F.; Williams, E. A.; Going, R. C.; Donahue, P. E.; Smith, J. F. J. Am. Chem. Soc. 1986, 108, 982.
45. van Vliet, M. R. P.; Buysingh, P.; van Koten, G.; Vrieze, K.; Koljic-Prodic, B.; Spek, A. L. Organometallics 1985, 4, 1701.
46. Benn, R.; Rufinska, A.; Janssen, E.; Lehmkuhl, H. Organometallics 1986, 5, 825.
47. Bougeard, P.; Mancini, M.; Sayer, B. G.; McGlinchey, M. J. Inorg. Chem. 1985, 24, 93.
48. Brevard, C.; Granger, P. Inorg. Chem. 1983, 22, 532.
49. Shehan, B. P.; Kony, M.; Brownlee, R. T. C.; O'Connor, M. J.; Wedd, A. G. J. Magn. Reson. 1985, 63, 343.
50. Green, J. C.; Grieves, R. A.; Mason, J. J. Chem. Soc. Dalton Trans. 1986, 1313.
51. Faller, J. W.; Whitmore, B. C. Organometallics 1986, 5, 752.
52. Tinkham, M. L.; Ellaboudy, A.; Dye, J. L.; Smith, P. B. J. Phys. Chem. 1986, 90, 14.
53. Spiccia, L.; Swaddle, T. W. J. Chem. Soc. Chem. Commun. 1985, 67.
54. Alberola, N.; Point, R. Int. J. Appl. Radiat. Isot. 1985, 36, 152.
55. Geraldes, C. F. G. C.; Sherry, A. D. J. Magn. Reson. 1986, 66, 274.
56. Doyle, G.; Heaton, B. T.; Occhiello, E. Organometallics 1985, 4, 1224.
57. Shimizu, T.; Hatano, M. Inorg. Chem. 1985, 24, 2003.
58. Pisaniello, D. L.; Helm, L.; Meier, P.; Merbach, A. E. J. Am. Chem. Soc. 1983, 105, 4528.
59. Jonas, J. Science 1982, 216, 1179.
60. Hasha, D. L.; Eguchi, T.; Jonas, J. J. Am. Chem. Soc. 1982, 104, 2290.
61. Batstone-Cunningham, R. L.; Dodgen, H. W.; Hunt, J. P.; Roundhill, D. M. J. Chem. Soc. Dalton Trans. 1983, 1473.
62. Roe, D. C. J. Magn. Reson. 1985, 63, 388.
63. Roe, D. C. Organometallics 1987, 6, 942.

64. Krusic, P. J.; Jones, D. J.; Roe, D. C. Organometallics 1986,
     5, 456.
65. Robert, J. M.; Evilia, R. F. J. Am. Chem. Soc. 1985, 107, 3733.

RECEIVED August 11, 1987

Chapter 8: Application 1

The T_1 Method
for the Detection
of Dihydrogen Complexes

Robert H. Crabtree, Douglas Hamilton, and Maryellen Lavin

Department of Chemistry, Yale University, New Haven, CT 06520

The temperature variation of the T_1 of the hydride
resonance of a polyhydride complex in the ^1H NMR can
be used to determine whether the structure is
classical, with terminal M-H bonds only, or nonclassical, containing one or more dihydrogen ligands
and to estimate the H-H distance.

The characterization of $W(H_2)(CO)_3(PCy_3)_2$ as a molecular hydrogen
complex, (1) as opposed to the classical dihydride formulation (2),
was based on neutron diffraction, the observation of a $\nu(H_2)$ in the
IR spectrum, and of a $^1J(H,D)$ in the ^1H NMR spectrum of the corresponding HD complex. (1) Similar methods have been used by others.
(2)

These methods are not always applicable. Neutron diffraction
requires suitable crystals. The IR band is more often absent than
present, especially in cationic complexes and in complexes in which
this band cannot gain intensity by coupling to a suitable intense
vibration, and the H-D coupling can only be seen if the complex is
not fluxional.
We needed a fast and reliable method which would work even on
unstable intermediates and other labile species in solution. We had
already shown (3) that the T_1 relaxation of hydrides in the ^1H NMR
is dominated by the dipole-dipole contribution. Since this depends
on the inverse sixth power of the distance between the interacting
nuclei, (4) we thought that the T_1 of a dihydrogen complex might be
unusually short, because the two hydrogens are so close together
(ca. 0.8A). (1) This distance should correspond to a relaxation
time of ca. 20ms.
Table I shows our T_1 results on some polyhydrides, obtained by
the conventional inversion-recovery method. As can be seen, they
fall into three groups. The first, with $T_1 > 300ms$, we regard as

0097–6156/87/0357–0223$06.00/0
© 1987 American Chemical Society

unambiguously classical, having terminal M-H bonds only. The
second, with $T_1 < 125ms$ are unambiguously nonclassical. We prefer
the formulations shown in Table I, in which there are one or two
dihydrogen ligands per metal because facile displacement only of the
H_2 ligands occurs with N_2 or phosphine to give classical species.
The third group has intermediate values of T_1.

The T_1 value also depends on the rotation of the molecule, as
measured by τ_c, the rotational correlation time. This will vary
according to the moment of inertia of the molecule and the viscosity
of the solvent. This variability, we thought, might account for the
intermediate values observed in Table I. On changing the temper-
ature the T_1 should go through a minimum when $\tau_c = 0.63/\omega$, where ω
is the NMR frequency. (4,6,9) The T_1 value at the minimum $T_1(min)$
is independent of the factors mentioned above and so is a useful
criterion for the structure determination. These values are shown
in Table II for a range of examples which have been studied to date.
The ambiguities apparent from the data of Table I are removed and we
are now able to assign the structures of all the species studied.

The $T_1(n,min)$ values so obtained reflect relaxation due to
dipole couplings (a) within the H_2 ligand and (b) between this
ligand and other dipolar nuclei in the complex. We have tried to
allow for (b) in eq. 2.

$$T_1(n,min,\ corrd.)^{-1} = T_1(n,min)^{-1} - T_1(c,min)^{-1} \qquad (2)$$

Once again, we assume $T_1(c,min)$ is 200ms. For the resulting $T_1(n,$
min, corrd.) we can obtain an H-H distance from eq. 3. (4,6,9) This
distance is also given in Table II. The numbers are only slightly

$$\{T_1(DD)\}^{-1} = 0.3\gamma^4 h^2 r^{-6}\{\tau_c/(1 + \omega^2\tau_c^2) + 4\tau_c/(1 + 4\omega^2\tau_c^2)\} \quad (3)$$

(γ = gyromagnetic ratio, h = Planck's constant, r = internuclear
distance, τ_c = rotational correlation time, ω = Larmor frequency)

affected by our choice of $T_1(c,min)$, e.g., a change of 10ms
typically changes r by 0.001A.

The difference in relaxation rates between the classical and
nonclassical group is so large that errors in measurement of T_1 (ca.
10%) are relatively unimportant. Paramagnetic samples would be
unsuitable, however, because all the relaxation times would be
short. This means that, so as not to be led astray by a paramagnetic
sample, it is necessary to verify that the T_1's of the ligand protons
are normal (3) even in a supposedly diamagnetic case. The presence
of a quadrupolar nucleus nearby (e.g., the metal) may affect the
hydride resonances, but it should not significantly affect the T_1,
as measured by inversion-recovery. (The presence of a quadrupolar
metal nucleus usually leads to rapid relaxation of the metal spin
states which will give rise to sharp M-H resonances. In unusual
situations there may be slow relaxation and broad M-H resonances may
be seen, but the T_1 of the M-H proton will not be affected. (9)) The
presence of other dipolar nuclei (e.g., ^{31}P) should not significantly
affect T_1 because of the much greater distance involved and in some
cases the lower values of the gyromagnetic ratio. The residual

Table I. The T_1 values of some hydrides

Complex	$T_1(ms)$[a]	$T_1(n)(ms)$[b]	Conditions	Refs.
$IrH_5(PCy_3)_2$	820		$CD_2Cl_2,-80°C$	5,6
$[IrH_2(H_2)_2(PCy_3)_2]^+$	48[c],73[d]	37	$CD_2Cl_2,-80°C$	5,6
$[IrH(H_2)(bq)L_2]^+$	30[c],390[d]	30	$CD_2Cl_2,-80°C$	6,7
$FeH_2(H_2)(PEtPh_2)_3$	24	12	toluene,$-70°C$	8
$RuH_2(H_2)(PPh_3)_3$	38	20	toluene,$-70°C$	8
$OsH_4(P\{p-tol\}_3)_3$	820		toluene,$-70°C$	8
$ReH_5(H_2)L_2$	78[e]	25	toluene,$-70°C$	9a
$ReH_5(H_2)dpe_2$	79[e]	25	$CD_2Cl_2,-70°C$	9a
" "	110	33	toluene,$-70°C$	9a
$Cp*ReH_6$	290,618		toluene,$70°C$	9a
ReH_5L_3	540		toluene,$-70°C$	9a
$MoH_4(PMePh_2)_4$	485		toluene,$-70°C$	9a
$MoH_3(H_2)(PMePh_2)_4^+$	44	18.5	$CD_2Cl_2,-70°C$	9a
$WH_4(PMePh_2)_4$	540		toluene,$-70°C$	9a
$[WH_5(PMePh_2)_4]^+$	148[f]		$CD_2Cl_2,-70°C$	9a
$WH_6(PMe_2Ph)_3$	166[f]		toluene,$-70°C$	9a
$H_2Fe(CO)_4$	3000		toluene,$-70°C$	9a

L = PPh_3, bq = 7,8-benzoquinolinate, Cy = cyclohexyl.
[a] by inversion-recovery, $\pm$ 10%. The values are somewhat solvent dependent and because of the τ_c term, temperature dependent. [b] calculated as shown in the text. [c] for the resonance assigned to the classical hydrides. [d] assigned to the nonclassical hydrides. [e] unpublished neutron diffraction data are said (10) to show disordered but classical structure for the $PPrPh_2$ complex. Perhaps difference is real and due to the change in phosphine, or there may be a change of structure on going to the solid state. [f] These relaxation times are intermediate between those which indicate an unambiguously classical structure (>300ms), and one which would suggest a nonclassical structure (<125ms). The data reported in Table II show that WH_6L_3 is classical but the $[WH_5L_4]^+$ cation has not yet been studied.

hydrides in deuteriated dihydrogen complexes show much smaller T_1's because of the lower gyromagnetic ratio of D. (5,6)

We hope to have shown that variable temperature T_1 measurements can be useful in the difficult problem of assigning structures to metal hydride complexes.

Acknowledgments

We thank the Petroleum Research Fund (R.H.C.) for funding, and Professors K. Zilm, J.W. Faller, J. Prestegard (Yale) and Dr. G. Hlatky (Exxon) for helpful comments, and W.A. Herrmann and J. Okuda for a gift of $Cp*ReH_6$.

Table II. Further T_1 data

Compound[a]	Temp(K)	T_1(min)	r(A)	structure
$CpRu(PPh_3)_2H$	238	300	-	classical
$[IrH_2(CO)_2(PPh_3)_2]^+$	210	245	-	classical
$IrH_5(PPh_3)_2$	<200	-	-	classical
$MoH_4(PMePh_2)_4$	250	165	-	classical
$ReH_7(PPh_3)_2$	200	55	0.90	nonclassical
$ReH_7(dpe)^b$	222	67	0.94	nonclassical
$ReH_7\{P(p-FC_6H_4)_3\}_2$	200	55	0.90	nonclassical
$ReH_7(PCy_3)_2$	<200	-	-	nonclassical
$ReH_8(PPh_3)^{-b}$	200	245	-	classical
$RuH_4(PPh_3)_3$	266	30	0.87	nonclassical
$WH_6(PMe_2Ph)_3$	235	181	-	classical

[a] solvents as in Table 1 [b] solvent: ethanol-d_6

Literature Cited

1. Kubas, G.J.; Ryan, R.R.; Swanson, B.I.; Vergamini, P.J.; Wasserman, H.J. J. Am. Chem. Soc., 1984, 106, 451.
2. Morris, R.H.; Sawyer, J.F.; Shiralian, M.; Zubkowski, J.D. J. Am. Chem. Soc. 1985, 107, 5581; Conroy-Lewis, F.M.; Simpson, S.J. Chem. Comm. 1986, 506.
3. Crabtree, R.H.; Segmuller, B.E.; Uriarte, R.J. Inorg. Chem., 1985, 24, 1949.
4. Pople, J.A.; Schneider, W.G.; Bernstein, H.J. High Resolution NMR; McGraw-Hill, New York, 1959.
5. Crabtree, R.H.; Lavin, M. Chem. Comm. 1985, 1661.
6. Crabtree, R.H.; Lavin, M.; Bonneviot, L. J. Am. Chem. Soc., 1986, 108, 4032.
7. Crabtree, R.H.; Lavin, M. Chem. Comm., 1985, 794.
8. Crabtree, R.H.; Hamilton, D.G. J. Am. Chem. Soc., 1986, 108, 3124.
9. a) Crabtree, R.H.; Hamilton, D.G., unpublished results; b) variable temperature T_1 studies have been reported (9c) but turning points were not seen; c) Jordan and Norton J. Am. Chem. Soc. 1982, 104, 1255.
10. Spencer, J.L.; Howard, J.A.K., personal communication.

RECEIVED August 11, 1987

Chapter 8: Application 2

A Convenient Method for Sealing and Opening NMR Tubes Under Air-Free Conditions

R. G. Bergman, J. M. Buchanan, W. D. McGhee, R. A. Periana, P. F. Seidler, M. K. Trost, and T. T. Wenzel

Department of Chemistry, University of California, Berkeley, CA 94720

Organometallic chemists have come to depend increasingly on NMR monitoring of reactions. With air-sensitive compounds, such experiments are often performed in flamed-sealed NMR tubes. This provides protection of the tube's contents from atmospheric oxygen, and allows volatile reagents to be added and sealed over the contents of the tube.

Use of Cajon fittings as a substitute for NMR tubes fused to glass joints.

One drawback to the use of sealed tubes is the necessity of fusing a glass joint to the top of the tube in order to connect it to a vacuum line. We have found that this can be avoided by use of a Cajon fitting. As shown in Figure 1, a Cajon Ultra Torr tube fitting (3/8" to 1/4" reducing union, model #SS-6-UT-6-4) attached to a 4 mm Kontes high vacuum Teflon stopcock (model #826500) can be used to attach an NMR tube to a vacuum source. Slightly larger O-rings (Viton, size 106) are used on the 1/4" end of the Cajon fitting since most NMR tubes are slightly smaller than 1/4" in diameter. Once a tube is attached to a vacuum line in this way and evacuated, volatile reaction components may be distilled in and the tube sealed off just as is done with a tube sealed to a glass joint.

Apparatus for cracking sealed NMR tubes and collecting volatile reaction products.

In the course of our research we have frequently found it necessary to recover not only the organometallic products from sealed NMR tube reactions, but also the volatile materials (when needed, for example, for quantitative analysis of gases by Toepler pumping and/or gas chromatography/mass spectrometric analysis). Often the organometallic complex is highly air sensitive and/or the volatile materials are gaseous (e.g. CH_4, CO, H_2), therefore requiring special handling. A method used in our group for recovering materials from sealed NMR tubes is described below.

Upon completion of the desired reaction the flame-sealed NMR tube is placed in the apparatus shown in Figure 2. The NMR tube must be scored and the score aligned with the stopcock A to insure a clean break of the tube (See Figure 2). The setup is then evacuated on a vacuum line while the contents of the NMR tube are cooled with liquid nitrogen (the tube is cooled to prevent rapid escape of the volatile materials upon exposure to the vacuum). Once evacuation of the apparatus is complete, stopcock B is closed and the sealed NMR tube is cracked by turning

0097–6156/87/0357–0227$06.00/0

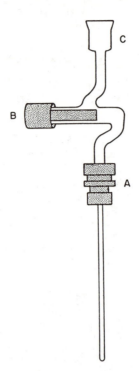

Figure 1. Apparatus for sealing NMR tubes: (A) cajon fitting;
(B) vacuum stopcock with teflon plug; (C) ground glass joint.

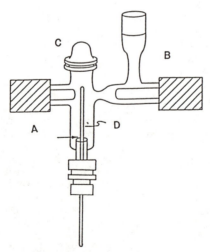

Figure 2. Apparatus for opening sealed NMR tubes: (A) teflon
plug; (B) vacuum stopcock with teflon plug: (C) Solv-Seal joint;
(D) sealed NMR tube. The arrow indicates the location of the
score on the NMR tube.

stopcock A. The contents of the tube can then be slowly thawed, enabling the volatile materials to be collected <u>via</u> vacuum transfer. Once the volatile materials are removed the entire apparatus can be transferred to a drybox for recovery of the organometallic products.

RECEIVED August 21, 1987

Chapter 9

New Methods for Acquiring IR Spectral Data in Organometallic Chemistry and Catalysis

Donald J. Darensbourg and Guy Gibson

Department of Chemistry, Texas A&M University, College Station, TX 77843

The advent of new infrared sampling accessories has allowed in situ study of organometallic systems under conditions that were previously not readily accessible. The techniques of Cylindrical Internal Reflectance (CIR) and Diffuse Reflectance spectroscopies are described herein. The CIR phenomenon was employed in three different apparatus. Two different high pressure CIR cells were used to study reactions homogeneously catalyzed by $[\mu-HW_2(CO)_{10}]^-$. Low temperature reactions of Mo and W complexes were studied using an ambient pressure CIR cell. The diffuse reflectance technique was employed to study powdered samples of Ru carbonyl complexes supported on Al_2O_3.

It would be a vast understatement to assert that infrared spectroscopy has played a major role in the characterization of organometallic derivatives, in particular metal carbonyls (1,2). Hence, a discussion of the role of infrared spectroscopy in organometallic chemistry and catalysis in our introductory comments will be avoided. Nevertheless, it is unquestionably true that the enhanced signal/noise available **via** FTIR has dramatically changed IR spectroscopy. As the technology has changed, the accessories for sampling have changed (3). These in turn have resulted in a significant impact on the type and quality of experiments that can be readily performed.

We wish to describe in this chapter some of the new and innovative infrared accessories commercially available to the experimental chemist. All of the infrared items described herein were obtained from Spectra Tech, Inc. (652 Glenbrook Road, P. O. Box 2190-G, Stanford, CT 06906, Ph. (800)243-9186). The conditions for which we have demonstrated the use of these devices involving in situ infrared spectroscopy include the following:

- High pressure and high temperature solution phase reactions carried out in both batch-mode or flow-through reactors.

0097–6156/87/0357–0230$06.00/0
© 1987 American Chemical Society

- Temperature controlled (high or low) solution phase reactions under gaseous atmospheres.

- Solid samples, high temperature under reactive gaseous atmospheres.

Cylindrical Internal Reflectance

A technique which has proven useful for our studies is that of cylindrical internal reflectance (CIR), coupled with a Fourier transform infrared spectrometer. In this study, an IBM-85 FTIR equipped with either a DTGS (deuterated triglycine sulfate) or MCT (mercury-cadmium-tellurium) detector was used. The infrared radiation is focused by concave mirrors onto the 45° conical ends of a transmitting crystal (Figure 1). The crystal may be made of any material which is optically transparent, has a high mechanical strength and high index of refraction, and is resistant to thermal shock and chemical attack. Suitable materials include ZnS, ZnSe, Si, Ge, and sapphire. As the beam passes through the crystal, it penetrates into the surrounding solution a distance of approximately 1.0-1.5µm. Therefore, ten reflections give a total pathlength of 10-15µm. The beam is then focused onto the detector. This is a relatively low throughput technique. The percent of the energy getting through the accessory without sample present is about 15% of the open beam throughput. However, good signal/noise ratios can be achieved using standard DTGS detectors, and improved ratios are achieved with MCT detectors. It is important to note that the internal reflection element and the input and output optics of the commercially available cells used in our efforts are maintained in a rigid, fixed position. This ensures that the pathlength is highly repeatable during a series of infrared measurements, a property pivotal to reliable quantitative results. Pathlength reproducibility is invariably a problem with conventional transmission cells used at high pressure and/or high temperatures.

External High Pressure Flow Cell. The principle of internal reflectance has been put to use in several different apparatus. The first of these used in our lab was a stainless steel high pressure (to 1500 psi) cell, the CIRCLE®, with a 25µl internal volume for the study of homogeneously catalyzed reactions (Figure 2). In this case, the cell was connected to a 300ml stainless steel high pressure reactor by means of 1/16" stainless steel tubing. Samples were periodically delivered from the reactor to the cell during catalysis with both being maintained at the same pressure. With suitable pumping, samples could be returned to the reactor, however in our instance samples were simply discarded after spectra were recorded. A high temperature jacket is available for this cell which can withstand temperatures up to 250°C.

   Figure 3 depicts FTIR spectra of the reaction of a CO_2/H_2 mixture and methanol at 125°C and 860psi to form methyl formate catalyzed by μ-H[$W_2(CO)_{10}$]$^-$ (4). The two strong peaks due to the starting metal carbonyl complex at 1940cm^{-1} and 1890cm^{-1} decrease while an attendant peak at 1980cm^{-1} due to W(CO)$_6$ increases in intensity. When all the starting complex has been consumed, a

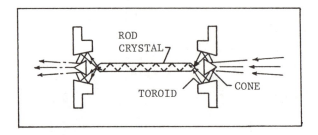

Figure 1. The cylindrical internal reflectance element.

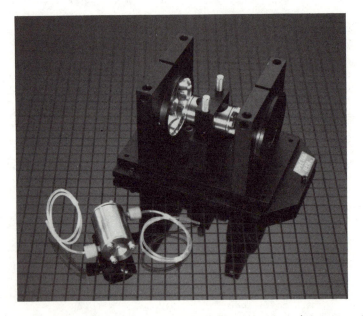

Figure 2. Micro CIRCLE high-pressure flow cell. (Photograph
courtesy of Spectra-Tech, Inc.)

weaker band is noted at about $1930 cm^{-1}$ (marked *) which is attributed to the metalloformate species $HCO_2W(CO)_5^-$. Based on the rate of methyl formate production (as evidenced by GC), the metalloformate derivative is described as an intermediate in the CO_2 reduction reaction. Definitive assignment of the peak at $1930 cm^{-1}$ to the $HCO_2W(CO)_5^-$ species was confirmed by an independent synthesis of this complex and its spectrum recorded under identical conditions (bold line in Figure 3).

A catalytic cycle based on these observations is shown in Scheme I. The decomposition of the starting material $HW_2(CO)_{10}^-$ to $HW(CO)_5^-$ and its subsequent reaction with CO_2 to yield $HCO_2W(CO)_5^-$ have been reported elsewhere (5,6). The presence of formic acid, which subsequently undergoes esterification to afford methyl formate, was detected by GC when the reaction was carried out in benzene.

in situ High Pressure Cell. Another means of obtaining high pressure FTIR spectra employs a modified Parr mini-reactor containing an imbedded internal reflectance crystal (Figure 4). This technique was first developed by Moser, et al. (7). The crystal is held in place by teflon retainers, and the reaction mixture is stirred by a magnetic impeller system. The reaction cell body is fabricated from 316 stainless steel and is intended to withstand pressures as high as 1000 psi and temperatures up to 250°C. The volume of the reactor is approximately 15ml. The reactor/cell is placed directly onto the optical bench of the FTIR instrument in our experiments. Potassium bromide windows are secured over the optical path openings in the instrument's sample compartment to prevent damage to the optical bench in case of reactor rupture. Alternatively, an external sampling bench is available with the CIR reaction cell (Figure 5) which offers protection of the IR instrument from hazardous experiments. Of course the use of this external bench results in additional loss of energy throughput (~50%).

This method of obtaining spectra is truly in situ spectroscopy. It is superior to the previously described high pressure method in that the reaction mixture need not be transferred to a separate cell, thus allowing cooling and undesirable reactions to occur. However, the external sampling technique offered by the flow reactor described above may be more applicable to large scale industrial use.

The in situ high pressure system was used to analyze the reaction of CO and methanol to form methyl formate, again catalyzed by μ-H$[W_2(CO)_{10}]^-$ (Figure 6) (8,9). A catalytic scheme has been proposed for this reaction whereby the catalyst is heterolytically cleaved under CO pressure, yielding $HW(CO)_5^-$ and $W(CO)_6$ (Scheme II). The metal hydride anion is protonated by methanol to give H_2 and the methoxide anion, which attacks a carbonyl ligand on $W(CO)_6$ to give a short-lived metalloester intermediate. This species is protonated by MeOH to give product and regenerate methoxide. The in situ IR of $W(CO)_6$ and OMe^- in THF at 500 psi CO pressure displays a three band pattern typical of $W(CO)_5X^-$, where X^- is presumably $C(O)OCH_3^-$. Although this metalloester intermediate was not isolated, the analogous more stable phenyl acyl complex has been more fully characterized. In addition, the $W(CO)_5OPh^-$ derivative has been the subject of an X-ray structure determination (10).

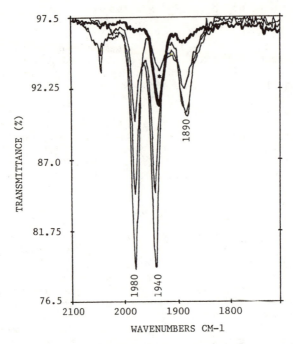

Figure 3. Infrared spectrum of operating catalyst system as a function of reaction time. Peak marked with asterisk due to $HCO_2W(CO)_5^-$.

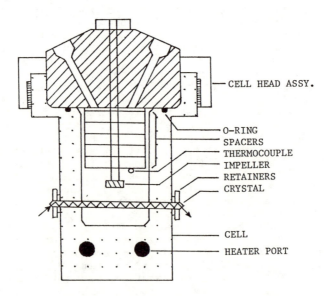

Figure 4. The CIR reaction cell.

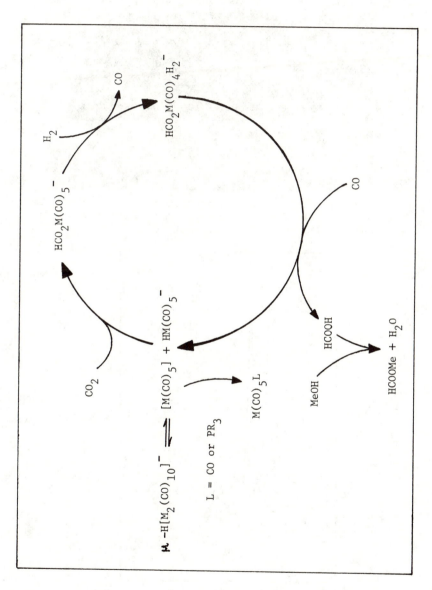

Scheme I.

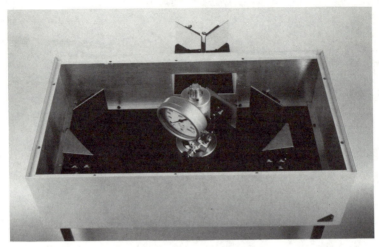

Figure 5. The external sampling bench, shown here with the CIR reaction cell installed at the center. (Photograph courtesy of Spectra-Tech, Inc.)

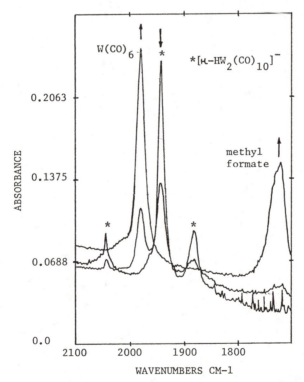

Figure 6. <u>in situ</u> infrared spectra as a function of time for the reaction of CO/MeOH with $[\mu\text{-HW}_2(CO)_{10}]^-$ as catalyst precursor.

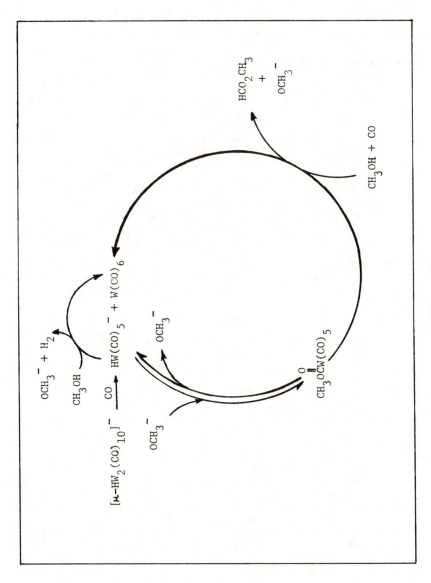

Scheme II.

Another catalytic system for which the in situ high pressure
cell proved useful was the reduction of cyclohexanone catalyzed by
$W(CO)_5(OAc)^-$ (11). A series of spectra were collected over a 24
hour period under 660 psi H_2 initial loading pressure (Figure 7).
The bands labeled (a) are due to the catalyst precursor
$W(CO)_5(OAc)^-$, while the (b) band is the carbonyl stretch of cyclohexanone. Upon heating to 125°C, new bands (c) due to
$W(CO)_4(OAc)\cdot THF^-$ appear. This assignment was based on a separate
photochemical experiment where $W(CO)_5(OAc)^-$ was irradiated in THF
solution. Further heating results in a buildup of $\mu-H[W_2(CO)_{10}]^-$
(peak (d)). The $W(CO)_4(OAc)\cdot THF^-$ species disappears upon cooling,
but the bridging hydride remains. Based on these observations, the
following reaction scheme was proposed (Scheme III).

Ambient Pressure Flow Cell. Cylindrical internal reflectance also
allows for in situ spectroscopy at atmospheric pressure. A pyrex
body (boat) cell for dynamic measurements is available in two
versions, micro (0.5ml) or macro (5.0ml), in which a crystal is
mounted inside a larger glass cylinder (Figure 8). Reaction
solutions can be pumped from a specially modified Kontes 50ml
Schlenk flask (Figure 9), through the cell, and back into the flask.
A glass coil made from 6 mm o.d. tubing was connected to the bottom
of the flask, and the tubing outlet was connected to the pump system
via a Kontes 12/5 ball joint. The flask can be cooled in a constant
temperature bath, and the coil on the bottom of the flask provides
for improved contact between solution and bath. The stopcock is
used to maintain an inert atmosphere by evacuation and backfilling
with nitrogen.

A gear pump with a maximum flow rate of 4.5 liters per minute
was connected to the cyclic system by means of polyethylene tubing.
The wetted parts of the pump are all either 316 stainless steel or
teflon, providing good chemical resistance. Polyethylene tubing was
chosen because of its relative inertness toward THF, the solvent in
most of our reactions. Peristaltic pumps were found unsuitable,
since no tubing was available which was both flexible enough for
such pumps and chemically resistant to THF. A typical flow rate was
20ml per second. This was as rapid as possible without causing
bubbles in the system. A fairly large volume of solution, about
50ml, was also necessary to prevent bubbles from forming. With this
system it was possible to maintain temperatures as low as -30°C
using a temperature bath filled with ethanol at -50°C. Lower temperatures are presumably attainable employing appropriate coolants,
e. g. dry ice/acetone. However, the atmosphere inside the optical
bench must be kept extremely dry to prevent condensation on the tips
of the crystal at these low temperatures.

This system was tested on a reaction which was first studied by
Poilblanc et al. (12); the protonation of $Mo(CO)_3(PMe_3)_3$ to yield a
seven-coordinate species $HMo(CO)_3(PMe_3)_3^+$. Figure 10A shows the
typical two band (1945cm^{-1} and 1841cm^{-1}) spectrum for a C_{3v} complex
such as $Mo(CO)_3(PMe_3)_3$. A 100mg sample of this complex was dissolved in 50ml of methylene chloride and the solution was cooled to
-15°C to help stabilize the protonated product. To the cooled
solution was added 1ml of CF_3COOH. Figure 10B shows the flow cell
spectrum of the protonated complex. A shift to higher frequencies
suggests protonation at the metal center. There is also a shift to

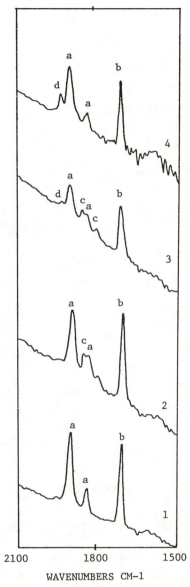

WAVENUMBERS CM-1

Figure 7. High pressure infrared spectrum in THF solution of
PPN$^+$W(CO)$_5$OAc$^-$ (a)/cyclohexanone (b) under 660 psig of H$_2$ initial
loading. (1) Initial spectrum 0.01 M PPN$^+$W(CO)$_5$(OAc)$^-$/0.2 M
cyclohexanone in THF at 25°C/660 psig of H$_2$. (2) Spectrum after
1 hr. at 125°C. The PPN$^+$W(CO)$_4$(OAc) THF$^-$ (labeled c) was
observed at 1861 and 1805cm^{-1}. (3) Spectrum after 5 hr. at
125°C. The band at 1939cm^{-1} is due to PPN$^+$µ-HW$_2$(CO)$_{10}$$^-$ (labeled
d). (4) Final spectrum after 24 hr. of reaction, cooled to 25°C.

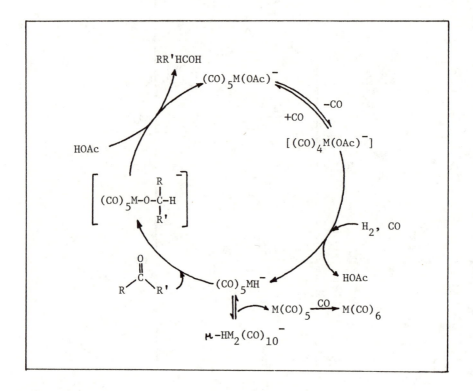

Scheme III.

Figure 8.   Ambient pressure glass (macro) flow cell.   (Photograph courtesy of Spectra-Tech, Inc.)

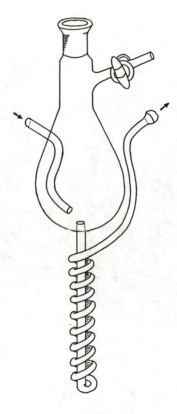

Figure 9.   Modified Schlenk flask for ambient pressure flow cell system.

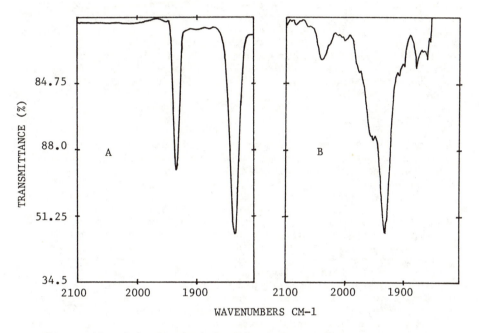

Figure 10.   (A) <u>fac</u>-W(CO)$_3$(PMe$_3$)$_3$ in THF at -15°C in flow cell.
(B) <u>after addition of CF$_3$COOH.</u>

a three band pattern ($2037cm^{-1}$, $1955cm^{-1}$, and $1930cm^{-1}$). If a C_{3v} arrangement of CO ligands was maintained, the lower frequency E band would be split into two equally intense peaks upon protonation. The different intensities of the peaks at $1955cm^{-1}$ and $1930cm^{-1}$ suggest a lower symmetry, C_s or C_{2v}.

The ability of the flow cell to do quantitative chemistry was demonstrated by the ^{13}CO enrichment of $W(CO)_5(O_2CCH_3)^-$. A solution of 0.5363g of $PPN^+W(CO)_5(O_2CCH_3)^-$ in 70ml of THF was cooled to 10°C, at which time the system was flushed with pure ^{13}CO. Figure 11 shows the disappearance of the peaks due to the all ^{12}CO labelled material and the growth of the ^{13}CO substituted tungsten derivative. A plot of the -ln(Abs) of the starting complex versus time illustrates first order kinetics in $W(CO)_5(O_2CCH_3)^-$. The ability to perform kinetic studies at low temperatures in situ under inert atmospheres, without heating up during the sampling process, should facilitate the obtaining of activation parameters for homogeneous solution reactions.

Diffuse Reflectance Cell

While cylindrical internal reflectance spectroscopy was used to analyze organometallic compounds in homogeneous solutions, the analysis of these compounds supported on inorganic supports (e.g., Al_2O_3, SiO_2, etc.) was carried out using a Diffuse Reflectance accessory, the COLLECTOR ^{TM} (Figure 12). This employs 4 flat mirrors and a curved mirror to focus the IR beam onto the flat surface of a powdered sample. The reflected beam is then focused onto the detector. The apparatus is fitted with a controlled environment chamber with which spectra can be obtained at elevated temperatures (to 600°C) and reduced or ambient pressures. The sample chamber can be conveniently loaded inside an inert atmosphere (argon) dry box.

Among the transition metals which catalyze the conversion of CO_2 to methane, ruthenium is probably one of the most studied (13). Anchoring $RuCl_3$ by impregnation on inorganic supports was reported in several instances to result in a catalyst for the reaction. More recently, it was shown that the use of ruthenium dodecacarbonyl in place of the mononuclear complex enhanced not only the conversion of CO_2, but also the selectivity toward CH_4 production (14,15). Furthermore, these cluster derived catalysts were found to have longer life and to give better conversion to methane at lower temperatures than their mononuclear counter parts.

Two catalysts were prepared by impregnating Al_2O_3 with $RuCl_3$ and $Ru_3(CO)_{12}$. Each sample was loaded into the controlled environment chamber (fitted with KBr windows), removed from the drybox, and secured to the diffuse reflectance accessory. Figure 13 shows the diffuse reflectance spectrum of $Ru_3(CO)_{12}/Al_2O_3$ prior to activation. No analogous spectrum for unactivated $RuCl_3/Al_2O_3$ exists because of the lack of carbonyl stretching absorptions. The $Ru_3(CO)_{12}$ catalyst was activated at 200°C under H_2, and the diffuse reflectance spectrum of the activated catalyst is shown in Figure 14A. The $RuCl_3$ catalyst was also activated at 200°C under H_2, then carbon monoxide was allowed to flow through the cell for several minutes, at which time the cell was evacuated and an infrared spectrum was obtained (Figure 14B). Comparison of the two spectra (14A and 14B) reveals that the ruthenium carbonyl surface species

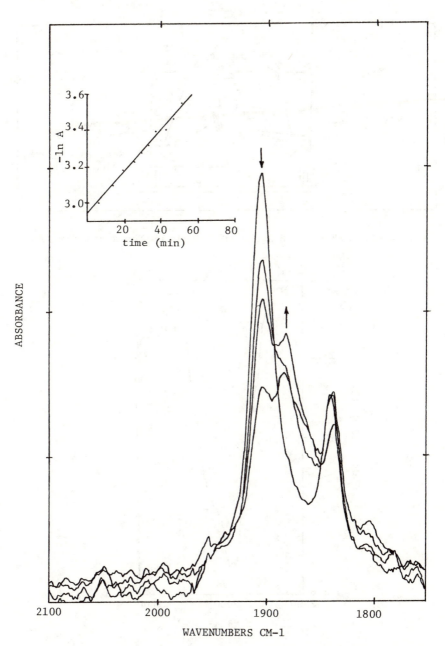

Figure 11. Reaction of W(CO)$_5$(OAc)$^-$ with ^{13}CO in THF at 10°C monitored in flow cell. Inset shows plot of -ln(abs) of the all ^{12}CO compound versus time in minutes.

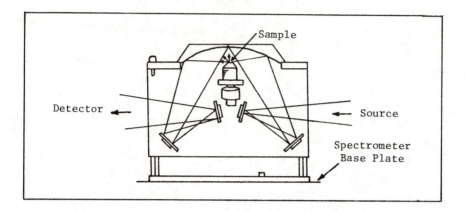

Figure 12. Schematic diagram of diffuse reflectance accessory.

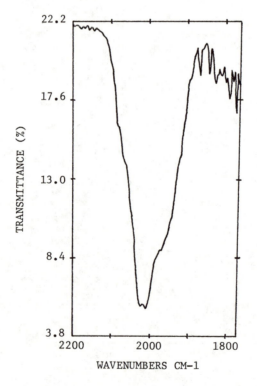

Figure 13. Diffuse reflectance infrared spectrum of $Ru_3(CO)_{12}$ supported on Al_2O_3, prior to activation.

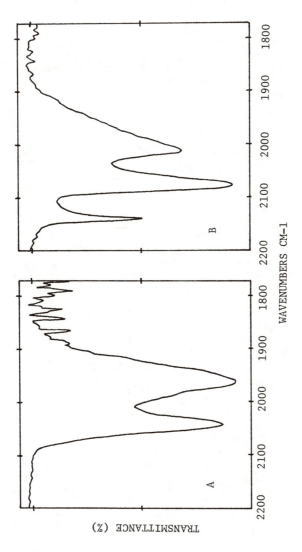

Figure 14. (A) Ru₃(CO)₁₂ on alumina (dehydroxylated at 150°C in vacuo for 24 hrs.) activated at 200°C under H₂. (B) RuCl₃ on alumina (dehydroxylated at 150°C in vacuo for 24 hrs.) activated at 200°C under H₂ followed by CO addition at 25°C.

are quite different for the two catalyst preparations. Similarly the reactivities of these two catalysts toward CO_2 hydrogenation to methane are quite different, with the $Ru_3(CO)_{12}$ derived catalyst being approximately 10 times more active. It should be noted that the spectrum obtained of activated $Ru_3(CO)_{12}$ on Al_2O_3 is essentially identical to that reported by Bell and coworkers on analogously treated wafered samples (16). We find that the ease with which the supported catalysts can be loaded in an inert atmosphere dry box, coupled with the lack of having to prepare wafered samples which transmit in the infrared and the ability to heat the sample under a flow of reactive gases, makes the COLLECTOR^{TM} accessory particularly suited to these research efforts.

Acknowledgments

We are most appreciative to Spectra Tech Inc., 652 Glenbrook Road, P. O. Box 2215, Stamford, CT 06906 for their generous support of our research efforts.

Literature Cited

1. Adams, D. M. Metal-Ligand and Related Vibrations; St. Martin's Press: New York, 1968.
2. Braterman, P. S. Metal Carbonyl Spectra; Academic Press: London, 1975.
3. See, for example, Spectra Tech Inc. brochure entitled "The Leading Edge in IR/ FTIR Accessories", Stamford, CT 06906, 1986.
4. Darensbourg, D. J.; Ovalles, C. J. Am. Chem. Soc. 1984, 106, 3750.
5. Darensbourg, D. J.; Rokicki, A.; Darensbourg, M. Y. J. Am. Chem. Soc. 1981, 103, 3223.
6. Darensbourg, D. J.; Rokicki, A. Organometallics 1982, 1, 1685.
7. Moser, W. R.; Cnossen, J. E.; Wang, A. W.; Krouse, S. A. J. Catal. 1985, 95, 21.
8. Darensbourg, D. J.; Ovalles, C.; Pala, M. J. Am. Chem. Soc. 1983, 105, 5937.
9. Darensbourg, D. J.; Gray, R. L.; Ovalles, C.; Pala, M. J. Molec. Catal. 1985, 29, 285.
10. Darensbourg, D. J.; Summers, K. M.; Rheingold, A. L. J. Am. Chem. Soc. 1987, 109, 290.
11. Tooley, P. A.; Ovalles, C.; Kao, S. C.; Darensbourg, D. J.; Darensbourg, M. Y. J. Am. Chem. Soc. 1986, 108, 5465.
12. Arabi, M. S.; Mathieu, R.; Poilblanc, R. J. Organomet. Chem. 1976, 104, 323.
13. For a recent review on the methanantion of CO_2 see: Darensbourg, D. J.; Bauch, C. G.; Ovalles, C. Rev. Inorg. Chem. 1985, 7, 315.
14. Ferbull, H. E.; Stanton, D. J.; McCowan, J. D.; Baird, M. C. Can. J. Chem. 1983, 61, 1306.
15. Darensbourg, D. J.; Ovalles, C. Inorg. Chem. 1986, 25, 1603.
16. Kuznetsov, V. L.; Bell, A. T.; Yermanov, Y. I. J. Catal. 1980, 65, 374.

RECEIVED September 14, 1987

Chapter 9: Application 1

A Simple Low-Temperature Solution IR Cell

Wolfdieter A. Schenk

Institut für Anorganische Chemie, Universität Würzburg, Am Hubland, D–8700 Würzburg, Federal Republic of Germany

In metal carbonyl chemistry, infrared spectroscopy continues to be the most often used physical method for product characterization. Occasionally it is desirable to obtain solution spectra at low temperature, e. g.,

- to investigate temperature-dependent equilibria,

- to record spectra of samples which decompose at ambient temperature in solution,

- to characterize unstable intermediates.

The cell itself is made out of two round pieces of brass with a diameter of 42 mm, into which rectangular openings (21 x 16 mm) are cut. The bottom part is 15 mm thick, a section 23 mm wide and 8 mm deep is cut as indicated in Fig. 1 to accommodate the windows. The top part is 8 mm thick and contains two bores which fit together with the holes of commercially available 38.5 x 22.0 x 4.0 mm solution cell windows. Luer-type syringe adaptors are soldered into these holes to provide an easy connection with metal or teflon tubing. The entire assembly is held together by four screws and spring washers as indicated in Fig. 1. Amalgamated lead gaskets should be preferred over the usual teflon to ensure good heat transfer. The cell windows should be made of water-resistant material such as calcium fluoride or Irtran.

The cooling block is cut out of a round piece of brass, 75 mm in diameter and 50 mm long. A concentric hole, 42.5 mm in diameter (a 0.5 mm gap between cell and cooling block is necessary to allow for thermal contraction) is cut out to depth of 45 mm; the remaining 5 mm are cut out to 30 mm diameter. On top of the block a metal box of approximate dimensions 50 x 45 x 25 mm is soldered. A small bore at the side of the block may serve as thermometer well. A rectangular piece of heavy-duty plastic with dimensions appropriate to slide into the sample holder of the spectrometer is attached to the back of the block (see Fig. 2 for details). Finally, two holes are drilled into the cooling block and plastic holder and fitted with brass tubings to provide a nitrogen purge.

0097–6156/87/0357–0249$06.00/0

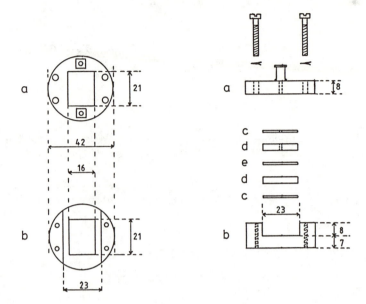

Figure 1. The cell: a) top part, b) bottom part, c) lead gasket, d) window, e) teflon spacer. All measures in mm.

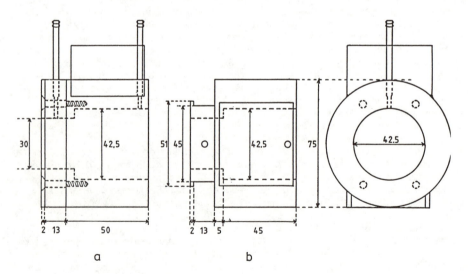

Figure 2. The cooling block: a) side view, b) top view, c) front view. All measures in mm.

To cool the entire assembly the metal box on top of the cooling block is filled with crushed dry ice and a few ml of isopropanol until the desired temperature is reached. A constant flow of N_2, precooled by passing it through a coiled copper tube immersed in liquid nitrogen, is passed through the block to prevent moisture from condensing onto the cell windows.

To obtain spectra of labile compounds a plastic syringe inserted into a suitable hole of a cold aluminium block is used to transfer the solution to the lower bore of the cell. A piece of teflon tubing connects the upper bore with a small vial to receive excess solution. This method has been successfully applied to obtain spectra of such labile compounds as $M(CO)_5(SO_2)$ (M = Cr, Mo, W)[1], cis-$W(CO)_4(PiPr_3)_2$[2], and cis-$W(CO)_4(PiPr_3)(C_2H_4)$[2]. Reactive intermediates can be observed by using two syringes containing the reactants and connecting them via small mixing chamber with the cell. The protonation of $[W(CO)_4(I)(PMe_3)]^-$[3] and the reactions of various olefin complexes[2,4] have been monitored by this method.

Solvent compensation is slightly more cumbersome than in ordinary work because all absorptions become sharper and increase in intensity with decreasing temperature. Use of a variable-pathlength reference cell alleviates most of the problems. The method of choice, however, is to run solvent spectra at various temperatures and use computer subtraction which is nowadays available even with medium-priced dispersive instruments.

A few precautions should be observed when working with this cell. Under conditions of high humidity it might be necessary to place a mask over the front of the cooling block with an opening just wide enough for the i.r. beam to pass. Also, it should be remembered that most single crystal window materials are sensitive to thermal shock. The cell should therefore be allowed to warm at least to 0°C before it is cleaned.

REFERENCES

1. Schenk, W. A.; Baumann, F. E.; Chem. Ber. <u>1982</u>, 115, 2615.

2. Schenk, W. A.; Müller, H., Z. Anorg. Allg. Chem. <u>1981</u>, 478, 205.

3. Schenk, W. A., unpublished.

4. Schenk, W. A.; Müller, H., Inorg. Chem. <u>1981</u>, 20, 6.

ACKNOWLEDGMENT

I thank Mr. W. Rüdling for his expert help in building this cell.

RECEIVED September 22, 1987

Chapter 9: Application 2

Quantum Yield Determination
by IR Spectrophotometry

Frederick R. Lemke and Clifford P. Kubiak

Department of Chemistry, Purdue University, West Lafayette, IN 47907

Quantum yields are often useful quantitative measures of the efficiency of photochemical processes. Among the more traditional methods of quantum yield determination is ferrioxalate actinometry by UV-vis spectrophotometry. Since many organometallic complexes contain strong infrared chromophores, e.g. C≡O, C≡NR, we describe herein a method for determining quantum yields in CaF_2 solution IR cells using a novel and simple actinometer: the photochemical disappearance of $Mn_2(CO)_{10}$ in neat CCl_4 to give $Mn(CO)_5Cl$ (<u>1</u>).

At a given wavelength of irradiation, the unknown quantum efficiency of a photoprocess, Φ_x, can be related to the known quantum efficiency, Φ_{Mn}, for the photoreaction of $Mn_2(CO)_{10}$ by Equation 1 (<u>2</u>).

$$\Phi_x = \Phi_{Mn} \frac{(\Delta A_x^{\nu_x}/\varepsilon_x)}{(\Delta A_{Mn}^{\nu_{CO}}/\varepsilon_{Mn})} \frac{\Delta t_{Mn}(1-10^{-A_{Mn}^{\lambda_{irr}}})}{\Delta t_x(1-10^{-A_x^{\lambda_{irr}}})} \qquad (1)$$

$\Delta A^{\nu} \equiv$ change in IR absorbance of the band at ν for the reacting species.

$\Delta t \equiv$ irradiation time.

$\varepsilon \equiv$ extinction coefficient of the IR band at ν.

$A^{\lambda_{irr}} \equiv$ UV-vis absorbance at the wavelength of irradiation.

$1-10^{-A^{\lambda_{irr}}} \equiv$ fraction of incident light absorbed by the sample at the wavelength of irradiation.

Note that if the experiment is carried out at concentrations and wavelengths such that $A^{\lambda_{irr}} > 1$ then the fraction of light absorbed

0097–6156/87/0357–0252$06.00/0

approaches unity and this latter term can be ignored. Equation 1 also is not reliable for large changes in $A^{\lambda}irr$. The idea is to work at optical densities as high as possible at the irradiated wavelengths, and to base quantum yields on the early phases of the reaction.

A solution of the photosensitive organometallic complex is prepared and freeze–pump–thaw degassed on a high–vacuum line. The solution concentration must be known so that the extinction coefficient, ε_x, of the IR band of the organometallic compound can be determined. The solution is transferred to a 0.05–0.2 mm pathlength CaF_2 solution IR cell. Exclusion of oxygen in photochemical experiments is extremely important. Solution transfer therefore should be performed in an inert atmosphere box. The extinction coefficient, ε_x, is determined by Beer's law from the concentration of complex and pathlength of your cell, before irradiation. The sample is photolyzed in the IR cell (CaF_2 transmits UV light to 120 nm (3)). [CAUTION: Eye protection to block out harmful UV irradiation is necessary. Regular safety goggles cutoff at ~ 290 nm. Special UV goggles with cutoff at ~ 366 nm are commercially available.] IR spectra are recorded over the course of photolysis, Δt_x, monitoring the absorbance changes, ΔA_x, of the IR chromophore of interest. Quantum yields are determined relative to the disappearance of $Mn_2(CO)_{10}$ in neat, degassed CCl_4. The disappearance of $Mn_2(CO)_{10}$, $\Delta A_{Mn}^{\nu CO}$, is conveniently monitored by the ν_{CO} band at 1970 cm^{-1} (ε_{Mn} (1970 cm^{-1}) = 4500 L mol^{-1} cm^{-1}). For samples irradiated at 313 nm, the quantum yield for disappearance of $Mn_2(CO)_{10}$, Φ_{Mn} = 0.48 (1). For samples irradiated at 366 nm, Φ_{Mn} = 0.41 (1). Clearly, IR actinometers for other regions of the UV–vis spectrum can be developed with accurate quantum yield data for photochemical reactions occurring at other wavelengths.

Literature Cited

1. Wrighton, M. S.; Ginley, D. S. J. Am. Chem. Soc. **1975**, 97, 2065.
2. Reinking, M. K.; Kullberg, M. L.; Cutler, A. R.; Kubiak, C. P. J. Am. Chem. Soc. **1985**, 107, 3517.
3. Gordon, A. J.; Ford, R. A. The Chemist's Companion: A Handbook of Practical Data, Techniques, and References; John Wiley and Sons: New York, 1972; p 180.

RECEIVED September 24, 1987

Chapter 9: Application 3

Spectrophotometric Cell for Inorganic and Organometallic Complexes Under Inert Atmospheric Conditions

Janet L. Marshall, Michael D. Hopkins, and Harry B. Gray

Arthur Amos Noyes Laboratory, California Institute of Technology, Pasadena, CA 91125

In the course of our research, we routinely measure UV-VIS absorption and emission spectra of inorganic and organometallic complexes under inert atmosphere conditions. In many cases, we are interested in monitoring the changes in the spectrum of a complex following the addition(s) of a given reagent. We have found that these experiments are facilitated by using the spectrophotometer cell shown in Figure 1.

This cell, which can be readily made by a professional glassblower, is designed for the preparation of solutions on a high vacuum line. It consists of a 10 mm pathlength square cuvette, of either Pyrex or quartz, and a small Pyrex bulb that is separated from the cuvette by a right angle valve. The cuvette of the cell shown in Figure 1 is a commercial fluorimeter cell with a 10 mm pathlength and attached graded seal tube (NSG Precision Cells, Type 63). Alternatively, the cuvette can be fashioned from 10 mm square tubing of either clear fused quartz (Wilmad Glass Company, WSQ-110, 20 mm I.D., 12 mm O.D.) or borosilicate glass (Wilmad Glass Company, WS-110-H, 10.0 mm I.D., 1.50 mm wall). The quartz cuvette constructed from the square tubing is attached to the round Pyrex tubing of the cell via a graded seal such as the Pyrex to Vycor glass graded seal (Corning 6466 from VWR Scientific). The small bulb used for the collection of the solvent and mixing of the solution is blown from heavy-walled Pyrex tubing to the desired diameter. The round tubing connecting the cuvette to the bulb is heavy-walled Pyrex tubing. The bulb is attached to the cuvette via a right angle Teflon vacuum valve (Kontes K-826610, size 4). Similarly, the cuvette is attached to a 24/40 female standard taper ground glass joint with the same type of right angle Teflon vacuum valve allowing the cell to be used readily on a vacuum line.

In practice, the inorganic or organometallic complex is loaded into the cell in an inert atmosphere box, if necessary, and the solution is prepared on a vacuum line by vacuum transfer of well-degassed solvent. After the desired spectrum is obtained, the solution is distilled into the Pyrex bulb by cooling the bulb in liquid nitrogen. The Teflon vacuum valve separating the cuvette from the pyrex bulb is closed and the additional reagent(s) can be added to the cell by removing the Teflon plug from the other vacuum valve. After addition of the reagent(s), which can be done in an inert atmosphere box, the cell is reevacuated on the vacuum line. Afterwards, the Teflon valve between the bulb and cuvette can be opened to allow mixing of the cell contents. This procedure can be repeated as often as desired.

0097–6156/87/0357–0254$06.00/0

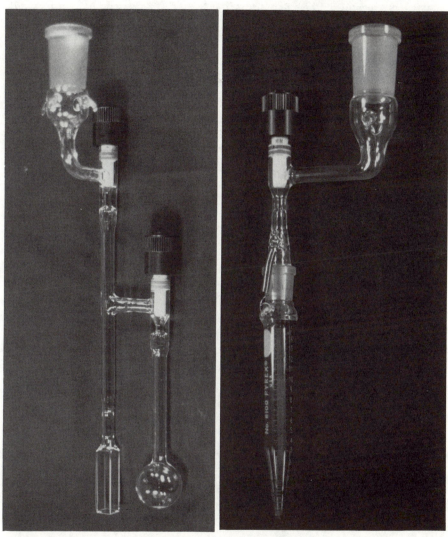

Figure 1 (left). Photograph of a spectrophotometer cell for measurements under inert atmosphere conditions. The cell pathlength is 10 mm.

Figure 2 (right). Photograph of the centrifuge-tube assembly for measurements of solvent volume.

For some experiments using this spectrophotometric cell, one may need to know the volume of solvent. This can be determined by first transferring the solvent into the modified graduated centrifuge tube shown in Figure 2 followed by vacuum transfer of the solvent into the bulb of the spectrophotometric cell. The centrifuge tube (15 mL capacity, Pyrex, Corning 8100 from VWR Scientific) is fitted with a 14/20 standard taper female ground glass joint. This tube is then used in conjunction with a right angle Teflon vacuum valve (Kontes K-826610, size 4) fitted with a 14/20 standard taper male ground glass joint and 24/40 standard taper female ground glass joint that can be attached to a vacuum line.

Acknowledgment

We thank Siegfried Jenner, Erich Siegel, and Gabor Faludi of the Caltech Glass Shop for their assistance. Our research in inorganic photochemistry is supported by the National Science Foundation. This is Contribution No. 7486 from the Arthur Amos Noyes Laboratory.

RECEIVED August 11, 1987

Chapter 10

Handling of Reactive Compounds for X-ray Structure Analysis

Håkon Hope

Department of Chemistry, University of California, Davis, CA 95616

Methods have been developed that allow simple handling
of highly reactive crystalline compounds in preparation
for x-ray analysis. The essence of the method consists
in protecting the crystals with a layer of hydrocarbon
oil during manipulation. Crystals can be prepared by
controlled atmosphere techniques, but thereafter all
handling can be carried out in the open. Routine use
of gas stream cooling (<140 K) during x-ray data collection has allowed long-term exposure with retention
of crystal integrity, without the use of protective
capillaries. Use of fast data acquisition methods
allows determination of new structures in 24 hrs, or
less, without use of other analytical techniques.

The most reliable and informative method of structure determination
is a single-crystal x-ray study. However, x-ray measurements can
require hours or days of radiation exposure, with attendant problems
in maintaining the chemical integrity of the sample under study.
Many important and intriguing compounds are highy reactive, and
cannot be exposed to the atmosphere without catastrophic conse-
quences. Sometimes even relatively stable compounds cause problems
when their crystals contain incorporated solvent which escapes when
the crystals are removed from the mother liquor. Loss of this
solvent can cause collapse of the crystal structure. Annoying
problems can also occur simply because the substance is hygroscopic.
    A commonly used answer to the incompatibility of chemical
properties and method of preferred structure determination has been
to protect the crystal by sealing it in a glass capillary. Many
otherwise inaccessible, but important structures have been determined
by use of this method. However, the procedure is far from simple,
and requires relatively elaborate equipment for maintaining an inert
atmosphere during manipulation of the specimen.
    In response to this situation a much simplified technique for
the handling of reactive materials has been developed. The essence
of the method consists in protecting the sample from the atmosphere

0097–6156/87/0357–0257$06.00/0

with a layer of inert, viscous oil, and in maintaining the crystal at low temperature during data collection.

The Case for Low Temperature

Cooling systems designed for routine use on x-ray diffractometers are almost exclusively based on the gas-stream principle, where the sample temperature is controlled with a stream of gas, usually dry nitrogen. Contemporary commercial low-temperature attachments allow reasonably easy temperature control in the range from near liquid nitrogen temperature to about room temperature.

Practically all crystals will be perfectly stable when kept in a nitrogen stream around 100 K. No solvent will escape, and reaction with oxygen or other atmospheric components is virtually impossible. This is the main reason for carrying out x-ray experiments on reactive samples at low temperature. However, consideration of the effect of cooling on atomic thermal parameters reveals another significant advantage. The intensity of a diffraction maximum is modified by a temperature factor, or damping factor, D, which in its simplest form is given by $D = \exp(-2B/(\sin\theta/\lambda)^2)$. The coefficient B is to a good approximation proportional to the Kelvin temperature. Common values of B for atoms in molecular crystals fall in the range 3 to 6 $Å^2$. Lowering of the temperature from 300 K to 100 K will reduce these to 1 to 2 $Å^2$. For $\sin\theta/\lambda$ around 0.5-0.6 $Å^{-1}$ this results in intensity enhancement by a factor of around 5, indicating that a fivefold increase in data collection speed can be realized without loss of data quality. For groups or ions such as ClO_4^- or PF_6^- that are notorious for large B values (15 $Å^2$ at 300 K is not uncommon), lowering the temperature can make the difference between a clean structural description and an intractable one.

The question of the effect of rapid cooling on the crystal is often raised, mainly with the concern that phase transitions or thermal stress may destroy the crystal. From experience with hundreds of compounds it can be stated with confidence that damage from cooling is rare. Perhaps about 1% of the crystals will undergo a phase transition under the conditions described here. Most of these cases can be handled by raising the temperature to a few degrees above the transition temperature. In practically no case is slow cooling of value; more frequently it will cause problems.

Setting up a low-temperature x-ray laboratory requires a certain investment of resources, and without a serious commitment the results are likely to be disappointing. Unfortunately, one cannot expect all cooling units to perform well as delivered by the vendor, although at least the more commonly sold models are basically sound. Up to this time the real responsibility for good performance has been left with the customer. Common problems are proneness to icing, excessive coolant consumption, and inadequate temperature stability. Solutions to the problems depend on model and make, and are beyond the scope of this book. Many unnecessary difficulties can be avoided by seeking advice from laboratories with successful low-temperature installations.

Preparation and Transfer of Crystals

The overall aim of the crystal handling technique is to keep operations as simple as possible, without undue risk to the sample. There are important practical and psychological reasons for this.

As is well enough known, it is rather common that a number of crystals must be examined on the diffractometer before a satisfactory specimen is found. Simple procedures that can be quickly executed will save much precious time. They are also more likely to result in a good crystal than are time-consuming, difficult methods, since there will be less resistance to discarding a marginal sample.

In general, it is desirable to prepare the crystals in a container with a relatively narrow mouth, and with the possibility of admitting an inert gas (typically N_2) through a separate inlet (a Schlenk tube is an example). In that way a reasonable overpressure of the gas is sufficient to prevent intrusion of atmospheric contaminants when the container is uncapped. With such a setup it is usually possible to carry out transfer of the crystals without use of a separate controlled atmosphere chamber.

Crystal growing techniques that are generally applicable to stable compounds can of course also be used with reactive compounds, provided adequate precautions are taken to avoid contamination. A large number of highly reactive crystals have been grown simply by placing a Schlenk tube with the appropriate solution (or reaction mixture) in a freezer at about -20°C.

Since more detailed accounts of preparative methods can be found elsewhere in this book, this section will not contain further discussion of that topic.

Retrieval of Crystals

General. As soon as crystals of appropriate size have been obtained it is time to transfer a specimen to the diffractometer. This is usually a two-step operation. First a sizable portion of the crystalline product is transferred to a flat-bottomed dish to which has been added a protective oil to a depth of a few mm, the amount depending on the expected reactivity of the compound. This laboratory has obtained excellent results with Paratone-N (Exxon), either neat, or mixed about 1:1 with mineral oil. For many compounds the resin part of epoxy glue has also worked well. The main requirements for the protective oil are fairly high viscosity (for slow diffusion of atmospheric gases and slow dissolution of the crystals), transparency to light (for convenience when working with a microscope), and inertness toward the compound in question.

Detailed Procedure. Two alternative, detailed procedures for transfer of crystals that decompose slowly or melt above room temperature are given.

First Procedure:
(a1) Add a 3 - 5 mm deep layer of the protective oil to a 5 - 10 cm petri dish.
(a2) Tilt the sample container so that a portion of the crystals moves to the neck of the flask, then tilt it back slowly so that these crystals remain in the neck. (At times it may be better to skip step (a2)).
(a3) Attach a gas supply hose to the flask, and apply a reasonable overpressure.

(a4) Open the flask, and with a spatula quickly scoop up a portion of the crystals, and stir them into the oil already prepared. The distance the sample is to be moved through air should be as short as practicable.

Second Procedure:
(b1) is the same as (a1).
(b2) Obtain a Pasteur pipette with a tip wide enough to suck up the crystals (1 - 2 mm) and flush it with inert gas.
(b3) is the same as (a3).
(b4) Open the flask, and withdraw a generous sample with the pipette. Avoid excessive amounts of solution.
(b5) Position the pipette with its opening in the oil and expel the crystals.

Sample Selection and Mounting. The sample should now be safe for a few minutes, and one can proceed to selection and mounting of a crystal. Selection of a suitable crystal is similar to procedures commonly used for chemically stable samples. A sowing needle held with a pin vice, and a sharp razor blade are useful tools. The binocular microscope should have a polarizing attachment to assist in judging quality of crystallinity.

Under the microscope a crystal with desirable attributes (size, apparent crystallinity) is selected and cleaned. The crystal is then removed from the oil by lifting it out with a prepared mounting fiber (see below for description), and immediately transferred to the cold stream on the diffractometer. This step can often be simplified if the crystal is moved toward the surface of the oil before one attempts to pick it up.

The distance between microscope and diffractometer should be short, no more than two or three meters. The transfer process can then be completed in about ten seconds. Because of the viscosity of the oil the crystal will be encased in a drop that will protect it during the transfer. At the low temperature in the cold stream the oil will become glass hard, and thus ensure a perfectly stable mounting.

Optical effects associated with the drop can interfere with the visual centering of the crystal, so that the amount of oil should be kept as small as possible, commensurate with the stability of the crystal.

A compound in crystalline form usually takes longer to suffer damage than one might be tempted to infer from its solution reactivity. Therefore it is good practice to determine the amount of air exposure a crystal will tolerate, and adjust the thickness of coating accordingly. Excess oil can be removed by sliding the crystal over a glass plate.

The Mounting Pin. In this laboratory the best results have been obtained with a specially designed mounting pin, illustrated in Figure 1. The pin is a hollow copper rod with a diameter corresponding to the adaptor hole in the goniometer head (normally 3 mm or 0.125 in). The tip of the pin is tapered down to 0.5 mm. The bore diameter near the tip is ~0.5 mm; the rest of the bore is 1.5 mm. In preparation for use the fine bore hole is filled with standard electronic solder (by capillary action). A glass fiber of

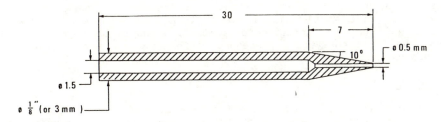

Figure 1. Dimensions of mounting pin. The overall length
will depend on diffractometer and gonoimeter head dimensions.
Material: Copper.

appropriate thickness is inserted into the molten solder, which is then allowed to cool. This results in a very stable assembly, with no tendency to drift with temperature changes. The glass fiber should extend 2-3 mm beyond the tip of the copper pin. It is critical that the glass does not extend outside the cold part of the cooling stream. The metal tip should be free of solder lumps, in order to prevent turbulence, which can cause icing. Copper, rather than brass, is used for its much better thermal conductivity.

Data Collection

In a paper from this laboratory it was shown that x-ray data collection can be substantially sped up without loss of significant information (1). With a Nicolet P2$_1$ diffractometer we collect 12,000 to 14,000 ω scan reflections in 24 h. Presumably, newer diffractometers can work at higher speeds. If crystals of reasonable quality are available, a 50-atom (non-H) structure should therefore be determined in less than 24 h. Once the crystal structure has been established, spectroscopic and other physical measurements can be planned and executed in a more rational manner than can be done with unknown samples. An x-ray structure determination should therefore normally be the first analytical procedure performed on a newly synthesized compound.

Limitations

The techniques described here are generally useful for compounds that melt near room temperature or higher. The author is developing a procedure for handling of low-melting crystals. Although initial experiments have been successful, the procedures are still fairly complicated, and further development is desirable.

Experience with the Method

The techniques described here were first developed in order to carry out x-ray data collection for phenyllithium etherate (2). More recently the structures of several extremely reactive organocuprates were determined in similar manner (3). A large number of other reactive compounds have also been prepared in the laboratory of Professor P.P. Power, and subsequently structurally characterized. The methods can be readily taught to new students, often requiring only a single demonstration for success. However, it should again be emphasized that good results depend on a properly working low-temperature apparatus, and a certain investment in equipment and acquisition of technique will therefore be required of those who want to take advantage of the methods described here.

Literature Cited

1. Hope, H.; Nichols, B. G. <u>Acta Cryst</u>. 1981, A<u>37</u>, 158-161.
2. Hope, H.; Power, P. P. <u>J. Am. Chem. Soc</u>. 1983, <u>105</u>, 5320-24.
3. Hope, H.; Olmstead, M. M.; Power, P. P.; Sandell, J.; Xu, X-J. <u>J. Am. Chem. Soc</u>. 1985, <u>107</u> 4337-38.

RECEIVED September 14, 1987

Chapter 10: Application 1

In Situ Crystal Growth for X-ray Diffraction Studies

Joseph C. Calabrese and Joel S. Miller

Experimental Station, E. I. du Pont de Nemours & Co., Wilmington, DE 19898

A technique for growing crystals *in situ* in a diffractometer enabling the X-ray structural determination of compounds that readily loose solvent upon harvesting is described.

Crystals of $[Fe(C_5Me_5)_2]^{\cdot+}[C_2(CN)_4]^{\cdot-}\cdot MeCN$ readily lose solvent upon collection and although appear nicely formed (ca. 1 x 1 x 10 mm) they diffract like powders. To obtain the crystal structure a method for *in situ* growth of the crystals in the diffractometer was utilized.[1,2]

A saturated solution of $[Fe(C_5Me_5)_2]^{\cdot+}[C_2(CN)_4]^{\cdot-}$ in MeCN was placed in a capillary and after initially sealing it with grease it was removed from the glove box[3] and flame sealed. The capillary was mounted into a diffractometer outfitted with FTS LT1[4] refrigeration device. The tube was then shock cooled via application of a cotton swab soaked with liquid nitrogen. A myriad of crystals formed[5]. The temperature of the capillary was warmed to just below the dissolution point of the crystals. A tiny hot wire[6] was then utilized to locally heat specific areas enabling the selective dissolution of undesired crystals. After reducing the number of crystals to 10-20 the temperature was lowered enabling these remaining crystals to grow larger. The warming, hot wire, cooling cycle was repeated until a single crystal remained which was then allowed to grow large enough to enable X-ray analysis. The temperature was then lowered to freeze the remaining supernate as an amorphous matrix.

[1] Previous papers using this technique include: Huffman, J. C. Ph. D. Thesis, Indiana University, **1970**; Calabrese, J. C.; Gains, D. F.; Hildebrandt, S. J.; Morris, J. H. *J. Am. Chem. Soc.*, **1976**, *98*, 5489; Rudman, R. "Low-Temperature X-Ray Diffraction: Apparatus and Techniques," Plenum Pub. Corp. **1976**.

[2] Miller, J. S., Calabrese, J. C., Rommelmann, H.; Chittipeddi, S. R.; Zhang, J. H.; Reiff, W. M. Epstein, A. J. *J. Am. Chem. Soc.*, **1987**, *109,* 000; Miller, J. S., Calabrese, J. C., Epstein, A. J.; Bigelow, R. W.; Zhang, J. H.; Reiff, W. M. *J. Chem. Soc., Chem. Commun.* **1986**, 1027.

[3] Due to air sensitivity a Vacumn Atmosphere Dry Box was used.

[4] FTS Systems, Inc. PO Box 158 Stone Ridge, NY 12484.

[5] A 30X Olympus stereomicroscope was used to monitor the growth of the crystals.

[6] A micro heating probe was made from a small round double bore ceramic tube (McDanel Refractory Porcelain Co., Beaver Falls, PA 15010) containing a nichrome wire powered by a variable power transformer.

RECEIVED September 1, 1987

Chapter 11

Photoelectron Spectroscopy

Experimental Characterization of Electronic Structure and Bonding in Organometallic Molecules

Dennis L. Lichtenberger, Glen Eugene Kellogg [1], and Louis S. K. Pang [2]

Laboratory for Electron Spectroscopy and Surface Analysis, Department of Chemistry, University of Arizona, Tucson, AZ 85721

The basic approach, instrumentation, sample requirements, and principles of interpretation for the photoelectron spectroscopy of organometallic molecules are
briefly described. The ionization information is
directly related to the formal oxidation states and d
electron counts of the metals, the actual effective
metal charges and charge potentials, the electronic
symmetry around the metal centers, and the individual
σ, π, etc. bonding capabilities of ligands and metal
fragments. This technique contributes to a level of
experimental characterization of organometallic molecules, beyond the usual three-dimensional structure
description, that is essential for interpretation of
many physical and chemical properties. Application
of the technique is illustrated with examples in
which the individual bonding capabilities of several
different ligands with a common metal fragment are
compared, and in which the bonding of different metal
fragments with a common ligand are compared.

The present level of understanding of organometallic chemistry and
catalysis is rooted in the detail and breadth of characterization
that is possible for organometallic molecules. An important key to
this knowledge is provided by characterization of the structural
arrangement of atoms in the molecule by crystal structure determina-
tions or other techniques. However, the understanding of the properties and behavior of a molecule, and the ability to extend this
understanding to new systems, requires knowledge of more than just
the particular three-dimensional arrangement of atoms. A further,
more detailed step in characterization of a molecule is to obtain
information on the electronic structure in terms of electron distributions, bonding, and stability. The characters of the bonds,
lone pairs, regions of charge build-up and depletion, local dipoles,
electronic perturbation effects, etc. define the physical and dynamic
properties of the molecule and its reactions with its environment.
Full characterization of a molecule must include its electronic
structural features in addition to the molecular structure. The
principles and models of organometallic bonding are some of the most

[1]Current address: Department of Chemistry, Northwestern University, Evanston, IL 60201
[2]Current address: University of Tasmania, GPO Box 252C, Hobart, Tasmania, 7001, Australia

important contributions to the modern renaissance of inorganic and
organometallic chemistry, and the techniques for obtaining this
information have experienced rapid development recently.(1)
Theoretical calculations of electronic structure have historically
had an important impact on organometallic chemistry, and theoretical
approaches and computational hardware continue to improve.
Experimental approaches directed toward electronic structure characterization are similarly essential.

In recent years photoelectron spectroscopy has been increasingly
utilized as an experimental tool to reveal electronic structure and
bonding features of solids, liquids, and gases.(2-12) Photoelectron
spectra offer well-defined and detailed experimental information
about electron richness, electron distributions, and the strength of
bonding interactions in or between molecules. The general electron
"richness" of a molecule may be correlated with its first ionization
potential (IP). A high ionization potential (binding energy) indicates relatively stable and inaccessible electrons, whereas a low
ionization potential indicates relatively unstable and available
electron density. These features are directly related to Lewis acid-
base and oxidation-reduction reactivity. The bonding within a molecule defines the molecular geometry and stability. Strong covalent
bonding interactions result in ionization bands with generally higher
IP's, whereas non-bonding and antibonding orbitals result in ionizations with generally lower IP's. The breadth of an ionization band
is also significant because it indicates the bonding character in the
ionized orbital and the extent of bond distance changes with the
ionization. Ionization associated with a strongly bonding orbital
results in a broad ionization envelope, whereas ionization associated
with a nonbonding orbital generally results in a narrow ionization
peak.

Principles such as these, along with the increasing library of
photoelectron ionizations, have led naturally to a variety of ioniza-
tion-structure-reactivity relationships. However, in order to obtain
meaningful information from photoelectron measurements it is important that the spectroscopic studies are adequately designed and
properly executed.   In particular, the bonding interactions are most
clearly revealed by the appropriate combination of spectroscopic
measurements (HeI UPS, HeII UPS, MgKα XPS, etc.) conducted on series
of free and complexed ligands and series of complexes with specifi-
cally useful interrelationships. Perhaps the most rewarding aspect
of this research is the exciting interactions between chemists in all
areas of research. The development of a complete bonding picture
must combine the skills of synthetic, theoretical, and physical
organometallic chemists. Syntheses of new classes of compounds
provides the opportunity to investigate the spectroscopy of new
bonding situations. Synthesis of a series of related molecules
provides the opportunity to investigate a particular bonding situation in a variety of environments. Theory aids in defining the
pertinent questions and principles under investigation.

This chapter is directed toward both new students and established researchers in organometallic chemistry whose expertise is not
in the area of photoelectron spectroscopy. It is recognized that
relatively few laboratories have direct access to photoelectron
instrumentation, but that many are interested in the information that
can be obtained. This chapter will briefly describe the photoelectron experiment and general sample requirements, and the
principles for understanding the information contained in the data.
The presentation of some "case studies" will illustrate to the reader

how photoelectron spectroscopy contributes to the experimental characterization of electronic structure.

The Experiment

Instrumentation. Conceptually, the photoelectron spectroscopy experiment is quite simple. Its roots are based in Einstein's photoelectric effect, in which light impinging on a metallic surface will ionize the surface if the energy of the photon of light exceeds the energy required to remove an electron from the metal. In gas phase photoelectron spectroscopy the excess energy of the ionizing photon is converted to the kinetic energy of the ejected electron. Thus the kinetic energy (E_k) of the ejected electron is the difference between the photon energy ($h\nu$) and the ionization energy, E_I (equivalently called the ionization potential, IP, or binding energy, BE).

$$E_k = h\nu - E_I$$

The photon energy of the light source is known, and measurement of the kinetic energy of the ejected electrons yields the ionization energies corresponding to the electron-occupied bound states. Kinetic energies are easily measured with an electrostatic analyzer as shown in Figure 1. The inner sphere is at a positive potential while the outer sphere is maintained at a negative potential. Electrons follow a curved path as they are repelled from the outer sphere potential and attracted by the inner sphere potential, and only electrons within a narrow range of kinetic energy will have the correct radius of curvature to pass through the exit slit to the detector at the end of the analyzer. The potentials on the spheres are scanned under computer control to observe different kinetic energy electrons. An analyzer vacuum of better than 10^{-4} Torr is necessary for the electron mean free path to be sufficiently long for electrons to reach the detector.

Several different monochromatic ionization sources are commonly used. Ultraviolet sources (photon energy generally less than 50 eV) such as HeI, HeII, or NeI provide information about valence electronic structure, and the experiment is labelled UPS for ultraviolet photoelectron spectroscopy. X-ray sources (photon energy generally greater than 1000 eV) such as MgKα, AlKα, etc., have sufficient energy to probe core ionizations, and the experiment is labelled XPS for x-ray photoelectron spectroscopy. Comparison of photoelectron spectra obtained with different energy photon sources provides information on the orbital character of the ionizations because each orbital type has a different ionization probability (cross-section) dependence on photon energy. Metal d orbital based ionizations show strong intensity with HeII excitation, but relatively weak intensity with HeI excitation. Ionizations of s and p orbitals localized on ligands comprised of carbon and other main group elements show the opposite trend. These changes are ideally suited for the study of organometallic molecules. HeI and HeII intensity comparisons are commonly used to reveal the metal and ligand character associated with the observed ionizations.

All components of the photoelectron spectroscopy instrumentation have continued to evolve over the last decade. New commercial sources for XPS with the anode at high positive potential have an order of magnitude improvement in photon flux over the older grounded anode designs. Analyzers with electron lens prefocusing are far superior to unmodified hemispherical, parallel plate, or cylindrical

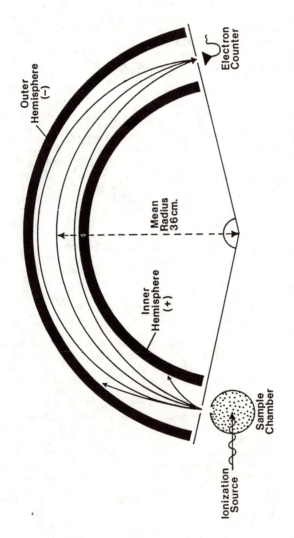

Figure 1. Schematic of hemispherical analyzer showing electrons with low kinetic energy hitting the inner sphere, electrons with high kinetic energy hitting the outer sphere, and electrons with kinetic energy in the correct range to reach the exit slit and detector.

mirror analyzers. Development of mini and microcomputers enables much more precise control of the analyzer focusing voltages and electron counting electronics.(13) The most important developments, however, have been in the areas of ultra-high vacuum (UHV) and clean sample handling/preparation in the UHV environment. These developments have made possible many experiments for investigating surfaces that were impossible a decade ago.

Although the photoelectron experiment is conceptually simple, utilization of photoelectron spectroscopy requires a substantial commitment of talented manpower and financial resources. The gas phase photoelectron experiment is particularly capricious because in the course of running the experiment the entire vacuum chamber is exposed to a partial pressure of the studied compound, thus degeneratively coating the electron optics, electron detector, ionization source, and pumping hardware. Diligent maintenance of surface conditions in the analyzer is necessary to retain sensitivity and resolution, and in many cases, depending on the instrument, a single experiment cannot be completed before loss of instrument performance. The best gas-phase photoelectron spectrometers for organometallic compounds are those least affected by these changes in surface conditions. Our instrument has a number of advantages in this regard. We utilize a relatively large 36 cm radius hemispherical analyzer with a 10 cm gap between the spheres (Figure 1). The design provides a large electron sampling angle and high electron throughput with good resolution (15-20 mV on the argon $2p_{3/2}$ line). The high sensitivity of the analyzer means that less sample is required, and the large surface area of the spectrometer means that more contamination is tolerated. In addition, the sample ionization chamber is separate from the analyzer chamber, and fast differential pumping systems are used to reduce the rate at which instrument surfaces become contaminated. Even with these advantages, a slow drift of the ionization energy scale is observed as the vapor pressure of the sample in the ionization chamber increases and deposition of sample on the slits and analyzer surfaces changes the electrostatics. Our high sensitivity and resolution allow us to use the sharp Ar $2p_{3/2}$ ionization reference line as a rapid internal lock (14) of the energy scale that maintains an energy scale constant to better than ± 0.003 eV. This internal lock is essential for observation of the detailed ionization band shapes and vibrational fine structure we are able to obtain.

Samples. Photoelectron spectroscopy has been used to study samples in the gas, liquid, and solid state. As with most forms of spectroscopy the sample type affects (severely) spectral resolution and sensitivity, and determines the amount and kind of information that can be obtained from a given study. The trade-off between resolution and sensitivity is most evident in the comparison of the photoelectron spectra of solids and gases. The high sensitivity from solid surfaces is compensated by solid state effects giving broader bands and a more complicated background. The relatively poor sensitivity of gas phase photoelectron spectroscopy is compensated by good resolution. A second major advantage of gas phase spectroscopy over solid state (surface, powder) spectroscopy is that the energy scale is easily and directly referenced to the vacuum level. We have recently made significant progress on reducing the problems of resolution, baseline, and energy referencing in the photoelectron spectra of solid organometallic samples on surfaces that promises to expand the future applications of ionization spectroscopy. However, at the

present time the principles of the photoelectron spectroscopy technique are best illustrated with studies of gas phase samples. The
remainder of this chapter will focus on gas phase results.

   Synthetic chemists who are interested in obtaining photoelectron
information on their systems commonly wish to know whether their
molecules are suitable for the experiment and how much sample is
required. Nearly any sample that can be sublimed can be studied in
the gas phase with photoelectron spectroscopy. Air and/or moisture
sensitivity is not an issue because sample transfers are performed
with inert atmosphere techniques and the actual experiment is performed in high vacuum. Another good indication that the gas phase
photoelectron experiment will be successful is provided by a normal
mass spectrum of the molecule showing the parent ion in high yield.
However, neither bulk sublimations nor mass spectroscopy exactly
match the conditions of the photoelectron experiment, and these
experiments can not be taken as absolute indicators for obtaining
suitable vapor pressure of the expected species in the gas phase for
photoelectron spectroscopy. In fact, quality photoelectron spectra
have been obtained for many species that do not hold together in bulk
sublimations or mass spectroscopy. The best experiment to answer the
question is simply to give a sample a try in the spectrometer. If
the sample happens to decompose in the spectrometer, the decomposition products are usually simpler molecules which are easily identified in the spectrum. For samples which are solids at room temperature, the experiment is carried out by slowly raising the temperature and thus the sample vapor pressure to obtain a series of
spectra until the sample is depleted. This is essentially a temperature programmed decomposition technique, and if no changes occur in
the spectra with time and temperature and no residue remains, then a
clean vaporization has taken place. Further confirmation of sample
integrity is provided by comparing the spectra of related series of
molecules. The amount of sample required depends on the behavior of
the molecule and the information that is desired. We have obtained
high quality (full valence region HeI) spectra on less than 10 mg of
sample. This amount is generally suitable for an initial experiment.
The HeII experiment is less sensitive and requires more sample. A
full HeI/HeII study might be accomplished with about 100 mg of a
well-behaved sample. The gas phase XPS experiment also involves low
sensitivity signal processing, and the core ionization energy measurement of each element in the molecule is repeated several times to
determine the confidence limits of the measurement. Complete gas
phase XPS studies may require several hundred milligrams of sample
with current instrumentation.

Ionization Energy Relationships

In a recent account we discussed in detail the information available
from ionization band shapes, intensities, splittings, and shifts in
the photoelectron spectra of transition metal compounds.(15) Here we
would like to give the reader a feeling for the relationships between
ionization energies and the traditional concepts of electronic structure and stability of organometallic molecules. First it must be emphasized that the ionization energy is simply a well-defined thermodynamic quantity that, unlike many other enthalpy and entropy quantities for organometallic molecules, can be precisely and relatively
easily measured. The fundamental significance of this quantity is
most easily appreciated by remembering the quantitative description
of the properties of atoms in freshman chemistry textbooks. An

important feature of the characterization of atoms is their valence ionization energies, which are then related to electronegativity scales, bonding schemes, and thermodynamic cycles. The characterizations of molecular ionization energies are significant for the same reasons.

The electronic structure factors that account for shifts in molecular ionization energies between two related molecules are closely similar to the factors that account for shifts in molecular orbital energies or eigenvalues. This is perhaps the primary reason that molecular photoelectron spectroscopy has attracted so much attention, although it should be stressed that there is considerably more value contained in the molecular ionizations than just this relationship. The valence ionization energies are often discussed as though they are experimentally measured molecular orbital energies (vide infra). The first consideration for the energy of a molecular orbital is the inherent stabilities of the constituent atomic orbitals and their contribution to the molecular orbital. The inherent stabilities of the outer valence electrons are related to the electronegativities of the atoms. The second factor contributing to the stability is the amount and kind of bonding character in the molecular orbital. Thirdly, the charge potential in the vicinity of the molecular orbital will shift the ionization energy by the electrostatic interaction.

The significance of these bonding and charge distribution features in the photoelectron spectra of transition metal complexes is apparent in the ligand additivity (16-19) and core-valence correlation (19-20) principles. These principles have recently been established to clarify the relationships between ionizations and molecular electronic structure, stability, and reactivity. Ligand additivity suggests that successive ligand replacements in transition metal complexes result in reproducible and systematic (additive) shifts of all valence and core ionization energies. The principle is clearly illustrated with the metal d ionization potentials of a series of $MA_{6-n}B_n$ complexes in which successive substitutions of ligands A with B may be compared. It has been shown that a metal-based ionization potential of any of these complexes ($E_n(i)$) is empirically predicted with the following relation:

$$E_n(i) = E_o + m(i) \, \Delta E_S(i) + n \, \Delta E_Q(i)$$

The first term represents the ionization potential of the unsubstituted MA_6 complex, and contains the inherent stability of the atomic orbitals comprising the metal d molecular orbital. The second term represents the bonding or overlap contribution to the ionization energy shift. The third term represents the change in charge potential contribution to the orbital ionization energy. The constant $\Delta E_S(i)$ is the difference in bonding stabilization between the two ligands A and B, and $\Delta E_Q(i)$ is the difference in charge potential provided by A and B. Thus a relative bonding stabilization and charge potential term can be characterized for each type of ligand bound to the transition metal center.

Ionization studies of core orbitals provide information complementary to valence ionization studies. While valence ionization energies include both contributions from charge distribution ($\Delta E_Q(i)$) and bonding or hyperconjugative ($\Delta E_S(i)$) effects, the core ionizations include the charge distribution contribution, $\Delta E_Q(k)$ (k representing a core orbital), but not the overlap bonding contribution. Core ionization energies can be used as an experimental indication of atomic charges (21-23) and comparison of a core shift with a valence shift can be used to separate out the bonding contribution to a valence ionization energy shift. Jolly's core-valence ionization correlation thus complements ligand additivity by relating core ionization energy shifts and valence localized orbital ionization energy shifts with a simple relation:

$$\Delta E_Q(i)/\Delta E_Q(k) = 0.8$$

Values substantially differing from 0.8 indicate a bonding (or antibonding) interaction providing an additional shift to the valence ionization.

    Ligand and metal replacements in series of related molecules represent perturbations of the basic electronic structure that allow determination of the relative E_0, ΔE_S, and ΔE_Q factors for the ionizations. Knowledge of the range of these factors for given substitutions also aids assignment of the ionizations and interpretation of the electron distribution and bonding of new molecules. For example, the effects of cyclopentadienyl ring methylation, e.g. cyclopentadienyl (Cp) vs. pentamethylcyclopentadienyl (Cp*), are utilized widely in organometallic chemistry. The electronic contributions to this effect have been characterized in detail by the photoelectron spectroscopy of sandwich and half-sandwich complexes.(24,25) The electronic effect of ring methylation for the cyclopentadienylmanganese tricarbonyls is shown in the spectra of Figure 2. HeI/HeII intensity comparisons indicate that the ionizations associated with the predominantly metal "t2g" electrons occur at lowest ionization energy on the right of the figure. These ionizations shift to lower energy with each ring methylation because of increasing charge donation to the metal from the ring. The slightly split ionization band at 9-10 eV correlates with the cyclopentadienyl e_1" ring π orbitals. The shift of this band with ring methylation is twice the shift of the metal band, again identifying it as having the predominant ring e_1" character. The slight splitting of this ionization feature follows from the symmetry reduction of the e_1" orbitals of free Cp^- when coordinated to the $Mn(CO)_3$ fragment. This shoulder is characteristic for these cyclopentadienyl ionizations of half-sandwich complexes. The large band growing in with ring methylation at around 11 eV corresponds to ionization of the e combination (C3v symmetry) of the C-H σ bonds of the ring methyls.(25) The overlap of this C-H σ bond symmetry orbital with the cyclopentadienyl e_1" orbital is a filled-filled orbital interaction that produces the large shift of the e_1"-based ionizations between Cp and Cp* analogues. Thus a large portion of the destabilization of the cyclopentadienyl valence π ionizations with methylation is due to orbital overlap effects rather than to inductive effects as commonly described. This is shown experimentally by the comparison of the core and valence ionization shifts.(25)

    The spectra of different molecules that are related by substitution of atoms down a group of the periodic table is another common example where ligand replacements can yield significant electronic structure information. For instance, two molecules may be the same except for the replacement of a chlorine atom by a bromine atom, or the replacement of an oxygen by a sulfur. The first effect observed from such a replacement is the altered stability of the ionizations caused by the different stability (electronegativity) of the atomic orbitals. Incorporation of a third row transition metal or heavy atoms such as iodine also introduces spin-orbit splittings that assist identification of the primary atomic character of the ionizations and the extent of mixing with other orbitals. These principles, along with a few others such as changes in ionization cross sections with source energy and the observation of vibrational fine structure mentioned in the experiment section, often are sufficient to overdetermine the assignment and interpretation of the electronic structure. That is, more than one experimental observation leads to a given conclusion. Theoretical calculations are then not necessary for drawing these conclusions from the experiment, and the experiment can be a true independent and unbiased test of the theories and models of organometallic chemistry.

The formal relationship between experimental ionization energies and theoretically calculated orbital eigenvalues is given by Koopmans' theorem.(26) Koopmans' theorem shows that the additional theoretical contributions of electron relaxation and electron correlation that appear as terms in ab initio orbital approximations are neglected in equating orbital eigenvalues to ionization energies. Both of these contributions are known to be large for transition metal species. Experimentally, electron relaxation energy differences between atoms within the same molecule or between molecules with differing atoms can also provide important bonding and electronic structure information. For example, the large difference in relaxation energy between Co and Rh explains many of the chemical differences that are manifested in the excited (positive ion) state, whereas ground state properties for analogous Co and Rh complexes are nearly identical.(27,28)

To illustrate to the reader the principles introduced in this section in the most direct and practical way, we present some case studies that compare the bonding properties of different ligands to the same metal center, and that compare the bonding properties of different metal functional groups to the same ligand. It is hoped that these experimental results will give the reader an appreciation of the types of detailed electronic structure information available from photoelectron spectroscopy.

Case Study: Bonding Capabilities of Ligands

This section will demonstrate the role of photoelectron spectroscopy in characterizing the relative bonding capabilities of ligands. The four ligands to be discussed are CO, PR_3, C_2H_4, and C_2R_2. These are quite general in organotransition metal chemistry, and knowledge of their bonding capabilities has been the focus of many investigations utilizing a variety of techniques. The valence photoelectron spectroscopy of the ligands as free molecules provides the first indication of the bonding capabilities of these ligands to metal complexes. The ionization energies of the highest occupied molecular orbitals for these species, which will act as the electron donors to the metals, are summarized in Table I. The order of electron richness as defined in the first section is $PMe_3 > C_2Et_2 > C_2H_4 > CO$. The donor abilities of these ligands will largely follow the same trend as long as overlap with the metal orbitals is similar. A more interesting feature of many organometallic ligands, however, is their π-acceptor capabilities. The valence spectra of the free ligands do not provide any direct information on the empty π orbitals. The π-acceptor capability of a ligand is experimentally characterized in photoelectron spectroscopy by observing the effect on the ionizations of occupied metal donor orbitals interacting with the ligand acceptor

Table I. Valence Ionization Energies for Ligands[†]

Ligand	Ionization	Ionization Potential, eV
CO	5σ	14.01
PMe_3	lone pair	8.58
C_2H_4	π bond	10.51
C_2Et_2	π bond	9.35

[†]Data for CO and C_2H_4 from reference 2, data for PMe_3 from reference 19, and data for C_2Et_2 from reference 36.

orbitals. Thus the π-acceptor abilities of two ligands can be compared if complexes are prepared in which those two ligands are bound to a common metal fragment or template.

The [CpMn(CO)$_2$] fragment displays many of the attractive features of a good template. Perhaps most important is the fact that a large number of substituent derivatives have been prepared and characterized. For instance, the spectra of CpMn(CO)$_3$, CpMn(CO)$_2$(PMe$_3$), CpMn(CO)$_2$(C$_2$H$_4$), and CpMn(CO)$_2$(C$_2$Et$_2$) can be compared to reveal the relative bonding capabilities of these ligands. The electronic structure and donor/acceptor capabilities of this fragment have been thoroughly characterized.(29-33) The cyclopentadienyl ring can be perturbed with ring methylations to shift valence ionizations, and the shift of the ring ionizations with ligand substitutions can reveal additional charge and π delocalization effects. CpMn(CO)$_2$L compounds may be considered pseudo-octahedral with the cyclopentadienyl ring occupying three coordination sites. A convenient coordinate system for understanding the orbital interactions in CpMn(CO)$_2$L places L along the z axis, with the x axis bisecting the two carbonyls as shown in Figure 3. The orbital contours shown are for calculated eigenvectors of the [CpMn(CO)]$_2$ fragment, i.e. not including L. The LUMO of the fragment is the 3a′ orbital, which is high in dz^2 character. This is the σ acceptor orbital of the metal for a ligand on the vacant z axis coordination site. The three highest occupied orbitals are the 1a′, a″, and 2a′ shown in Figure 3. The 2a′ and a″ both possess π symmetry with respect to the z-axis coordination site, and both can act as π donors. The 2a′ is largely dxz in character. The a″ is the HOMO (dyz). Ligands with single π donor orbitals (e.g. olefins) tend to align for interaction with the a″.(29,30) The 1a′ (dx^2-y^2) interacts with the two in-plane carbonyls and has a primarily δ symmetry interaction with the vacant coordination site. The shift in ionization energy of the 1a′ orbital with ligand substitution on the z axis largely reflects the total change in charge potential.

Valence Ionization Data and Band Assignments. Close up HeI (5-11 eV) spectra for Cp*Mn(CO)$_3$ (25), Cp*Mn(CO)$_2$(PMe$_3$) (34), Cp*Mn(CO)$_2$(C$_2$H$_4$) (35), and Cp*Mn(CO)$_2$(C$_2$Et$_2$) (36) are displayed in Figure 4. The "parent" tricarbonyl complex displays ionizations which correlate with the metal valence d orbitals and the cyclopentadienyl ring e$_1$″ orbitals, as discussed earlier. The low energy ionization band represents the predominantly metal d^6 electrons and the very small broadening of this ionization envelope in the tricarbonyl complex supports the view of these molecules as pseudo-octahedral with a nearly degenerate "t2g" set of metal d electrons. The additional valence ionizations of the PMe$_3$, C$_2$H$_4$, and C$_2$Et$_2$ complexes (9-10 eV) correlate with the donor orbitals of these ligands. The donor

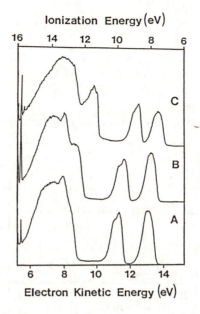

Figure 2. HeI full (6-16 eV) spectra for (A) $(\eta^5\text{-}C_5H_5)Mn(CO)_3$, (B) $(\eta^5\text{-}C_5H_4CH_3)Mn(CO)_3$, and (C) $_3(\eta^5\text{-}C_5(CH_3)_5)Mn(CO)_3$.

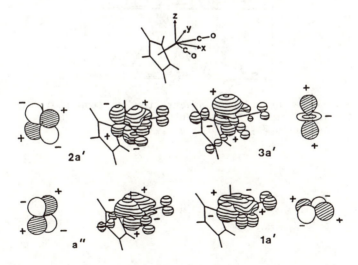

Figure 3. 3-dimensional orbital contours of the [$(\eta^5\text{-}C_5H_5)Mn(CO)_2$] fragment with respect to a vacant ligand on the z-axis, with the dominant metal orbital contribution shown next to the contours. The 3a' correlates with the dz^2 σ acceptor, the a" correlates with the dyz π donor, the 2a' correlates with the dxz π donor, and the 1a' correlates with the $dx^2\text{-}y^2$ δ donor.

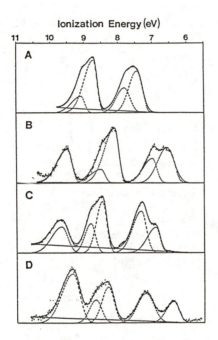

Figure 4. HeI closeup spectra for (A) Cp*Mn(CO)$_3$, (B) Cp*Mn(CO)$_2$(PMe$_3$), (C) Cp*Mn(CO)$_2$(C$_2$H$_4$), and (D) Cp*Mn(CO)$_2$(C$_2$Et$_2$). Cp* = η^5-C$_5$(CH$_3$)$_5$.

orbital of CO is more stable and ionizes in the envelope from 12-16 eV. Note that using the pentamethylcyclopentadienyl (Cp*) ligand rather than the unsubstituted Cp ligand enables the clean separation of the Cp e_1" ionizations from ligand donor ionizations of these complexes. (35,36)

The valence ionizations of $CpMn(CO)_3$, $Cp*Mn(CO)_3$, $Cp*Mn(CO)_2(PMe_3)$, $Cp*Mn(CO)_2(C_2H_4)$, and $Cp*Mn(CO)_2(C_2Et_2)$ are characterized by their vertical ionization potentials, band shapes, and relative areas in Table II. Independent half-widths are used to model the high and low energy sides of the asymmetric Gaussian peaks. Valence ionization bands are broader on the high binding energy side because of the increasing slope of the potential well to the higher excited vibrational levels of the postive ion.(31) The differences in ionization potentials are the most informative data, and will be discussed in detail in the Analysis section.

The deeper core ionization energies of these complexes are obtained with the x-ray ionization source. Table III lists the core ionization data for $CpMn(CO)_3$, $Cp*Mn(CO)_3$, $CpMn(CO)_2(PMe_3)$, $CpMn(CO)_2(C_2H_4)$, and $CpMn(CO)_2(C_2Me_2)$. In this case the emphasis is placed on the unmethylated cyclopentadienyl ring compounds because of cleaner separation of the ionizations in the carbon 1s ionization region, where the carbon 1s ionizations of the carbonyls are separated from the carbon 1s ionizations of the cyclopentadienyl ring and PMe_3. For reference, the data for both $CpMn(CO)_3$ and $Cp*Mn(CO)_3$ have been listed in Tables II and III. Core spectra are displayed in Figure 5 for the manganese $2p_{3/2}$, oxygen 1s, carbon 1s, and phosphorus 2p ionization regions of $CpMn(CO)_2(PMe_3)$.

Analysis. With this general assignment of the ionizations of these complexes, it is now possible to examine the information that these ionizations provide for the relative bonding capabilities of the different ligands. The most dramatic differences are in the metal ionization regions. The t2g set of metal orbitals in the pseudo-octahedral tricarbonyl (Figure 4A) are only slightly split (mostly as a consequence of the local C_{3v} symmetry of the $Mn(CO)_3$ portion of the molecule) into an a + e set. These are in the relative area ratio of 1:2. Interactions between the third (z-axis) CO and the $[CpMn(CO)_2]$ fragments a" (HOMO) and 2a" donor orbitals are shown below.

It should be noted for comparison with the bonding of olefins and alkynes that both metal orbitals are stabilized by donation into empty carbonyl acceptor orbitals.

When the incoming ligand is a phosphine, the ligand "acceptor" orbitals are the very high-lying phosphorus d levels, and the interaction is expected to be weak. In fact, the weakness of this interaction allows experimental inference of the electronic structure of the $[CpMn(CO)_2]$ fragment. Figure 6 is an ionization correlation diagram (similar to a molecular orbital energy diagram, but with

Table II. Valence Ionization Energies of [Cp]Mn(CO)$_2$L Complexes

Complex	Ionization Label	Ionization Energy, eV	Half-Widths, eV W_H	W_L	Relative Area
CpMn(CO)$_3$	M1	7.97	0.65	0.38	1.00
	M2	8.33	0.65	0.38	0.58
	Cp1	9.84	0.58	0.33	0.91
	Cp2	10.25	0.58	0.33	0.46
Cp*Mn(CO)$_3$	M1	7.46	0.55	0.39	1.00
	M2	7.82	0.55	0.39	0.58
	Cp1	8.72	0.67	0.25	1.26
	Cp2	9.09	0.67	0.25	0.37
Cp*Mn(CO)$_2$(PMe$_3$)	M1	6.60	0.63	0.50	1.00
	M2	7.01	0.63	0.30	0.58
	Cp1	8.11	0.73	0.38	1.40
	Cp2	8.51	0.73	0.28	0.35
	P1	9.49	0.74	0.30	0.88
Cp*Mn(CO)$_2$(C$_2$H$_4$)	M1	6.94	0.61	0.32	1.00
	M2	7.34	0.61	0.32	1.60
	Cp1	8.48	0.42	0.28	1.55
	Cp2	8.81	0.42	0.28	0.88
	L1	9.67	0.71	0.40	1.19
Cp*Mn(CO)$_2$(C$_2$Et$_2$)	M1	6.40	0.53	0.34	1.00
	M2	7.22	0.58	0.53	1.77
	Cp1	8.31	0.46	0.37	1.62
	Cp2	8.67	0.46	0.37	1.05
	L1	9.35	0.77	0.46	3.31

[†]Data for CpMn(CO)$_3$ and Cp*Mn(CO)$_3$ from reference 25, data for Cp*Mn(CO)$_2$(PMe$_3$) from reference 34, data for Cp*Mn(CO)$_2$(C$_2$H$_4$) from reference 35, and data for Cp*Mn(CO)$_2$(C$_2$Et$_2$) from reference 36.

Table III. Core Ionizations for [Cp]Mn(CO)$_2$L Complexes

Complex	Ionization	Ionization Potential, eV	FWHM, eV
CpMn(CO)$_3$	Mn 3d$_{5/2}$	646.76	1.3
	O 1s	538.85	1.6
	C 1s - Cp	291.76	1.3
	C 1s - CO	292.76	1.3
Cp*Mn(CO)$_3$	Mn 3d$_{5/2}$	646.08	1.3
	O 1s	538.37	1.6
	C 1s - Cp	290.71	1.6
	C 1s - CO	292.35	1.3
CpMn(CO)$_2$(PMe$_3$)	Mn 3d$_{5/2}$	645.36(4)	1.3
	O 1s	537.53(4)	1.6
	C 1s - Cp/PMe$_3$	290.27(3)	1.6
	C 1s - CO	291.67(6)	1.6
	P 2p$_{3/2}$	136.54(4)	1.8
CpMn(CO)$_2$(C$_2$H$_4$)	Mn 3d$_{5/2}$	645.93(4)	1.5
	O 1s	528.19(2)	1.7
	C 1s - Cp	290.67(8)	1.7
	C 1s - CO	292.23(8)	1.7
	C 1s - C$_2$H$_4$	289.9(2)	1.7
CpMn(CO)$_2$(C$_2$Me$_2$)	Mn 3d$_{5/2}$	645.92(7)	1.5
	O 1s	537.97(3)	1.7
	C 1s - Cp	290.6(1)	1.8
	C 1s - CO	292.0(1)	1.8
	C 1s - C$_2$Me$_2$	289.6(2)	1.8

[†]Data for CpMn(CO)$_3$ and Cp*Mn(CO)$_3$ from reference 25, data for CpMn(CO)$_2$(PMe$_3$) from reference 34, data for CpMn(CO)$_2$(C$_2$H$_4$) and CpMn(CO)$_2$(C$_2$Me$_2$) from reference 36.

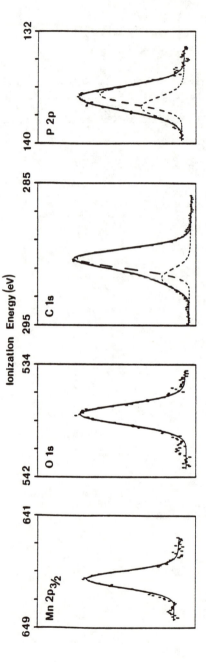

Figure 5.　Core XPS ionizations for $(\eta^5\text{-}C_5H_5)Mn(CO)_2(PMe_3)$.　The Mn $2p_{3/2}$, oxygen 1s, carbon 1s, and phosphorus 2p regions are displayed.　The carbon 1s feature is deconvolved with two peaks because two distinctive forms of carbon (CO and Cp/PMe₃) are observed in this complex.　The phosphorus 2p ionization is spin-orbit split into a $^2P_{3/2}$ and $^2P_{1/2}$ doublet.

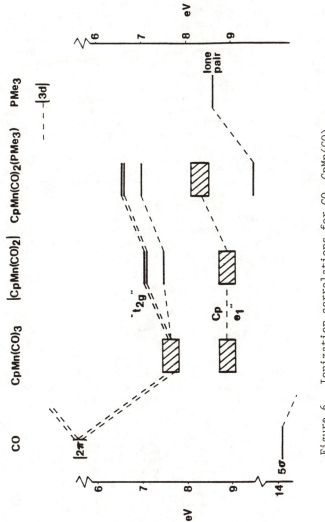

Figure 6. Ionization correlations for CO, CpMn(CO)$_3$, CpMn(CO)$_2$(PMe$_3$), and PMe$_3$. Levels for the [CpMn(CO)$_2$] fragment are estimated based on the ionization correlations.

experimental ionization energies) for [CpMn(CO)$_2$] interacting with CO and PMe$_3$ to form CpMn(CO)$_3$ and CpMn(CO)$_2$(PMe$_3$). The splitting of the occupied metal levels of [CpMn(CO)$_2$] is the same as observed for the phosphine complex since there is no significant phosphine overlap interaction with these orbitals. The ionization energy shifts of the fragment metal levels due to the charge potential provided by the phosphine ligand is directly reflected in the ionization energy of the 1a'. In addition to the 1a' ionization energy, the core ioniza-tion data of Table III and the valence cyclopentadienyl π orbital ionization potentials all experimentally show the charge potential difference between CO and PMe$_3$. For example, the difference in core (Mn 2p$_{3/2}$) ionization potentials between CpMn(CO)$_3$ and CpMn(CO)$_2$(PMe$_3$) of -1.40 eV indicates about a 0.10-0.15 electron (22) potential difference at the metal center. This is partly the result of the loss of π backbonding stabilization concomitant with the replacement of CO with PMe$_3$, and partly due to the electron richness (and σ donation) of PMe$_3$ compared to CO. The π backbonding dif-ference between CO and PMe$_3$ is shown by the splitting of the pre-dominantly metal valence ionization bands. The higher ionization energy band of the phosphine complex corresponds to the one metal orbital backbonding with two carbonyls, and the lower energy band corresponds to the two metal orbitals interacting with only one carbonyl and the phosphine. This splitting is 0.41 eV, less than one-third of the core shift.

Ethylene provides a different situation because it has only one π acceptor orbital. The ethylene's preferred orientation has this acceptor orbital interacting with the [CpMn(CO)$_2$] a" orbital. There

is no acceptor orbital on ethylene that can interact with the 2a'. Thus only one of the two donor orbitals is stabilized by the ethylene and the metal ionization profile is represented by a 1:2 pattern of peaks. The left side of Figure 7 shows this interaction in an ioni-zation correlation between the [CpMn(CO)$_2$] fragment and the ethylene. Note that the degeneracy in the second metal ionization indicates that the one ethylene π acceptor orbital is as effective as a single CO π acceptor orbital at stabilizing a given metal-based ionization. Also interesting is that the splitting between the two metal ioniza-tions is 0.40 eV, nearly identical to that observed in CpMn(CO)$_2$(PMe$_3$). This underscores the statement that PMe$_3$ has negli-gible π acceptor capability. Core ionizations show that C$_2$H$_4$ is intermediate between CO and PMe$_3$ in terms of the charge potential at the metal.

The alkyne ligand has a π acceptor orbital similar to C$_2$H$_4$ that serves to stabilize the 1a". The alkyne is especially interesting in comparison with carbonyls and olefins because of its different interac-tion with the other potential metal π donor orbital, the 2a'. The 2a', rather than being stabilized by a π* orbital as in the case of carbonyl or having no ligand interaction as in the case of olefins, now experiences an interaction with a filled π orbital on the alkyne.

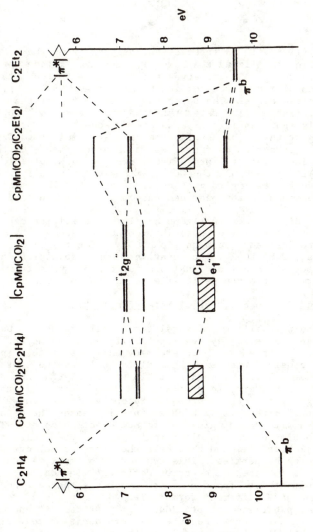

Figure 7. Ionization correlations for C_2H_4, $CpMn(CO)_2(C_2H_4)$, $CpMn(CO)_2(C_2Et_2)$, and C_2Et_2. Note that C_2H_4 has one π acceptor orbital and is only able to stabilize one metal level. C_2Et_2 has two π acceptors, but only one is oriented to accept from the metal. The other metal orbital is destabilized by a filled-filled interaction with one of the filled alkyne π orbitals.

This leads to a filled-filled interaction, destabilizing the metal 2a' orbital, while stabilizing the lower lying alkyne π (see the right side of Figure 7). One observes a 1:2 pattern in the metal ionization region of the alkyne complex with a large 0.82 eV splitting between the two bands. The degeneracy between two of the valence metal ionizations indicates that the one π acceptor orbital of this alkyne is again as effective as one CO acceptor. Core ionizations suggest that C_2R_2 donor properties are similar to C_2H_4. A filled-filled interaction does not result in a net change in the charge distribution. The cyclopentadienyl ring in this molecule is not an innocent ligand because the Cp e_1'' mixes substantially with the alkyne π levels, leading to π delocalization.(36) This mixing largely occurs because of the near degeneracy of the alkyne and Cp π orbitals.

In summary, differences in ligand bonding capabilities can be examined with valence and core photoelectron spectroscopy. By using the same template fragment (such as [CpMn(CO)₂]) for the different ligands (CO, PMe₃, C_2H_4, C_2R_2) the relative σ donor abilities and π acceptor abilities of each individual orbital can be quantitatively described in terms of the energetic effects on the molecule.

Case Study: Bonding Capabilities of Metal Fragments

The relative bonding abilities of _metal_ functional groups can be obtained in an analogous way by keeping the ligand constant and changing the metal fragment. There are many examples of this approach. First, consider complexes that differ only by the central transition metal. If the two metals are in the same group but differ greatly in relaxation energy (e.g. first and second row),(27) the relaxation energy effect can be probed. Secondly, it was shown above that replacement of carbonyls with phosphines increases the metal electron richness without changing the formal electron count. Thus comparing the bonding of two metal fragments (e.g. [CpMn(CO)₂] and [CpMn(CO)(PMe₃)]) with a ligand L reveals the effect of electron richness on bonding properties. This type of experiment is especially interesting in cases of electron deficient ligands.

Here we would like to focus on metal fragments that are nearly isostructural but have different formal electron counts and are bound to ligands with appropriate acceptor/donor properties. Alkynes are classic ligands in this regard because they can donate 2 to 4 electrons.(37) Another well-known example is the nitrosyl ligand.(23,38) Having thoroughly examined d⁶ CpMn(CO)₂(C_2R_2), an appropriate companion study is to investigate d⁴ CpV(CO)₂(C_2R_2).(36) The [CpV(CO)₂] fragment orbitals are essentially the same as those of [CpMn(CO)₂] but with two fewer electrons. The removal of one pair of electrons from the metal should eliminate the destabilizing filled-filled interaction between the metal and alkyne that pushed the HOMO to its very low binding energy in Cp*Mn(CO)₂(C_2Et_2). This also means that vanadium is inherently electron deficient and the alkyne must be a four electron donor for an 18 electron complex.

Valence Ionization Data and Band Assignments. Figure 8 shows the HeI full spectra for $CpV(CO)_2(C_2H_2)$ and $CpV(CO)_2(C_2Et_2)$.(36) A close up HeI/HeII fit of the data for $CpV(CO)_2(C_2H_2)$ is displayed in Figure 9 and the analytical band deconvolution data for both complexes are listed in Table IV. The HeI/HeII comparison for the acetylene complex clearly indicates that the first ionization band (7 eV) is metal d in character. The four bands between 9 and 10.5 eV are ligand (alkyne and cyclopentadienyl). Because the two higher energy bands of these four bands shifts more than the lower energy pair between the C_2H_2 and C_2Et_2 complexes (Figure 8), the two higher energy ionizations are correlated primarily with the alkyne π levels and the lower energy bands are correlated with the cyclopentadienyl ring. Two other points worth noting are the small shoulder on the high ionization energy side of the first ionization peak, and the fairly large shift and broad profile of the second ionization band. Both of these factors have important implications for understanding the electronic structure of this complex.

Table IV. Valence Ionization Energies of $CpV(CO)_2(C_2R_2)$ Complexes

Complex	Ionization	Ionization Potential, eV	Half-Widths, eV W_H	W_L	Relative Intensity HeI	HeII
$CpV(CO)_2(C_2H_2)$	M1	7.02	0.31	0.23	0.8	1.3
	M1'	7.30	0.31	0.23	0.2	0.4
	M2	8.32	0.58	0.51	1.0	1.0
	Cp1	9.23	0.44	0.28	1.7	1.9
	Cp2	9.58	0.44	0.28	0.9	0.9
	L1	9.95	0.75	0.40	1.4	1.7
	L2	10.38	0.75	0.40	1.4	2.2
$CpV(CO)_2(C_2Et_2)$	M1	6.77	0.30	0.22	1.0	
	M1'	7.03	0.30	0.22		
	M2	7.81	0.54	0.46	0.9	

M1' is a vibrational fine structure shoulder on M1 (see text).

The Cp1, Cp2, L1, and L2 bands overlap severely in the 8.5-9.5 eV region of the $CpV(CO)_2(C_2Et_2)$ spectrum and were not deconvoluted.

Analysis. The HeI/HeII intensity analysis shows that the low energy band at around 7 eV is a predominantly metal ionization. This band also shows a small shoulder on the high binding energy side due to vibrational fine structure associated with the CO stretching frequency. We have found this structure in the ionizations of other metal-carbonyl complexes when the metal orbital is backbonding symmetrically to the carbonyls.(14,17) In this case the metal dx^2-y^2 (1a') is backbonding symmetrically to the carbonyls. Ionization of this orbital removes electron density with some CO π^* character, and the resultant change in CO equilibrium bond length in the positive ion produces the short progression in the CO stretching vibrational mode. The splitting of the shoulder from the main (vertical)

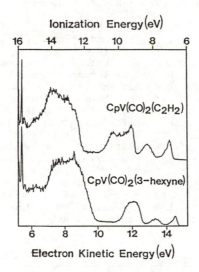

Figure 8. HeI full spectra for CpV(CO)$_2$(C$_2$H$_2$) and CpV(CO)$_2$(C$_2$Et$_2$).

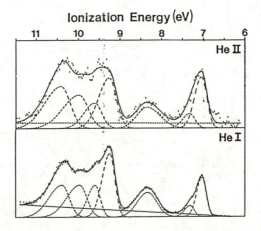

Figure 9. HeI/HeII comparison for CpV(CO)$_2$(C$_2$H$_2$). The 7 eV band is deconvolved with two asymmetric Gaussians because of vibrational fine structure (CO stretch) on this strongly backbonding metal ionization. The 8.5 eV band shows substantial ligand character.

component of the ionization is 0.26 eV (2100 cm^{-1}). This is slightly higher than the CO vibrational frequency of 1900-1980 cm^{-1} observed in the IR spectrum of the neutral complex, as expected for removal of electron density that is antibonding between the carbon and oxygen atoms.

The information from the second ionization band (8 eV) is especially important. In comparison to the analogous d^6 manganese complex this vanadium complex is formally considered to be d^4, and the second ionization band would be ascribed to the second pair of metal electrons. However, the ligand-like HeI/HeII response of the second ionization band, the large shift of its ionization energy between the C$_2$H$_2$ and C$_2$Et$_2$ complexes, and the broad profile of the band, which is characteristic of bonding character, all indicate substantial alkyne character in this ionization. Both π donor orbitals of the alkyne have been identified with the ionizations above 10 eV. The 8 eV ionization must then correlate with the second metal donor orbital which is substantially delocalized into the alkyne π^*. The third metal orbital from the t2g parentage (that accounted for the filled-filled metal-alkyne interaction in the Mn case) is now a good acceptor orbital with respect to the alkyne π. The metal σ acceptor is still the 3a' (dz^2). Thus, there is a net four electron donation from the alkyne ligand to the metal.(36)

A large amount of electron density has also been back-accepted by the alkyne as evidenced by the relatively low ionization energies of its π orbitals (L1 and L2): 9.95 and 10.38 eV vs. 11.43 eV in free C$_2$H$_2$. The fact that the alkyne has donated 4 electrons to the metal allows it to be a good π^* acceptor ligand. Also indicative of this synergism is the large ligand character of the second ionization band, which would formally be assigned to the metal if it is considered to be d^4. Actually, the formal oxidation state of the vanadium is more correctly viewed as +III (d^2) since only one pair of electrons can clearly be assigned to the metal in the ionization spectrum (the first band). The important point is that the ionizations are actually more consistent with the metallocyclopropene description of the bonding. The three σ bonds of the metallocyclopropene are two metal-carbon bonds and one carbon-carbon bond. Two strong metal-carbon bonds are observed in the ionizations. The symmetric combination of the metal-carbon bonds correlates with the donation of the alkyne π bond (10-11 eV) to the metal 3a' and the antisymmetric combination of the metal-carbon bonds is indicated by the second ionization at 8 eV. The carbon-carbon σ bond is much more stable and ionizes in the region of the other carbon-carbon σ bonds of the cyclopentadienyl ring (above 12 eV). Finally, the carbon-carbon π bond of the metallocyclopropene is observed near 10 eV, and is able to donate into the empty metal level. The large distortion in the alkyne R-C-C angle (140°) and the increased carbon-carbon bond distance observed in the crystal structure of this complex (36) reflect the metallocyclopropene nature of bonding.

Synopsis

These cases illustrate just some of the kinds of information provided by photoelectron spectroscopy for the characterization of organometallic molecules. Much of this information is extremely basic to the understanding of organometallic molecules, but is generally not available from other experimental techniques. For example, the *formal* oxidation state of the metal is indicated by the valence ionizations, where the number of ionizations identified with high metal character gives the number of formally metal d electrons. The effective charge at the metal center, however, is a physical characteristic that is different from the formal oxidation state. Two metals with the same formal oxidation states in two different complexes will have different effective charges dependent on the electron donation and acceptance properties of their ligands. The combination of valence and core photoelectron spectroscopy provides a measure of the charge potential at the metal center. The splitting of the valence metal ionizations also shows the electronic symmetry of the ligands around the metal center. The shifts of the individual valence ionizations measure the relative bonding capabilities of individual ligand and metal complex orbitals.

These principles extend to the study of more complex organometallic molecules and clusters, and current improvements in gas phase and surface photoelectron instrumentation promises to make the use of these techniques more routine than has been the case in the past. Photoelectron spectroscopy is taking a place alongside crystallography, theoretical calculations, and other common techniques for characterization of new organometallic molecules.

Acknowledgments

This work has been supported by the U. S. Department of Energy (Division of Chemical Sciences, Office of Basic Energy Sciences. Office of Energy Research), the National Science Foundation, the Petroleum Research Fund, Research Corporation, and the Materials Characterization Program, Department of Chemistry, University of Arizona.

Literature Cited

1. Hoffmann, R. Science, 1981, 211, 995.
2. Turner, D.W.; Baker, C.; Baker, A.D.; Brundle, C.R. Molecular Photoelectron Spectroscopy; Wiley: New York, 1983.
3. Ghosh, P.K. Introduction to Photoelectron Spectroscopy; Wiley: New York, 1983.
4. Brundle, C.R.; Baker, A.D. Electron Spectroscopy: Theory, Techniques, and Applications; Academic: London. Volume 1, 1977, Volume 2, 1978.
5. Siegbahn, K.; Nordling, C.; Johansson, G.; Hedman, J.; Heden, P.F.; Hamrin, K.; Gelius, U.; Bergmark, T.; Werme, L.O.; Manne, R.; Baer, T. ESCA Applied to Free Molecules; North Holland: Amsterdam, 1969.
6. Hendrickson, D.N. in Physical Methods in Chemistry; Drago, R.S., Ed.; Saunders: Philadelphia, 1977.
7. Rabalais, J.W. Principles of Ultraviolet Photoelectron Spectroscopy; Wiley: New York, 1977.
8. Briggs, D. Handbook of X-ray and Ultraviolet Photoelectron Spectroscopy; Heyden: London, 1977.
9. Shirley, D.A. Electron Spectroscopy; American Elsevier: New York, 1972.
10. Green, J.C. Struct. Bonding (Berlin) 1981, 43, 37-112.
11. Cowley, A.H. Prog. Inorg. Chem. 1979, 26, 45-160.
12. Solomon, E.I. Comments on Inorg. Chem. 1984, 3, 227-320.
13. Lichtenberger, D.L.; Kellogg, G.E.; Kristofski, J.G.; Page, D.; Turner, S.; Klinger, G.; Lorenzen, J. Rev. Sci. Instrum., 1986, 57, 2366.
14. Hubbard, J.L.; Lichtenberger, D.L. J. Am. Chem. Soc., 1982, 104, 2132.

15. Lichtenberger, D.L. Kellogg, G.E. <u>Acc. Chem. Res.</u>, in press.
16. Bursten, B.E. <u>J. Am. Chem. Soc.</u> 1982, <u>104</u>, 1299-1304.
17. Bursten, B.E.; Darensbourg, D.J.; Kellogg, G.E.; Lichtenberger, D.L. <u>Inorg. Chem.</u> 1984, <u>23</u>, 4361-65.
18. Bancroft, G.M.; Dignard-Bailey, L.; Puddephatt, R.J. <u>Inorg. Chem.</u> 1984, <u>23</u>, 2369-70.
19. Lichtenberger, D.L.; Kellogg, G.E.; Landis, G.H. <u>J. Chem. Phys.</u> 1985, <u>83</u>, 2759-68.
20. Jolly, W.L. <u>Acc. Chem. Res.</u> 1983, <u>16</u>, 370-6.
21. Stucky, G.D.; Matthews, D.A.; Hedman, J.; Klasson, M.; Nordling, C. <u>J. Am. Chem. Soc.</u>, 1972, <u>94</u>, 8009.
22. Jolly, S.L.; Perry, W.B. <u>Inorg. Chem.</u> 1974, <u>13</u>, 2686.
23. Hubbard, J.L.; Lichtenberger, D.L. <u>Inorg. Chem.</u> 1980, <u>19</u>, 1388-90.
24. Cauletti, C.; Green, J.C.;Kelly, M.R.; Powell, P.; Van Tilborg, J.; Robbins, J.; Smart, J. <u>J. Electron Spectrosc. Relat. Phenom.</u> 1980, <u>19</u>, 327-53.
25. Calabro, D.C.; Hubbard, J.L.; Blevins, C.H. II; Campbell, A.C.; Lichtenberger, D.L. <u>J. Am. Chem. Soc.</u> 1981, 103, 6839-46.
26. Koopmans, T. <u>Physica</u> (Utrecht) 1934, <u>1</u>, 104-13.
27. Lichtenberger, D.L.; Calabro, D.C.; Kellogg, G.E. <u>Organometallics</u> 1984, <u>3</u>, 1623-30.
28. Lichtenberger, D.L.; Blevins, C.H. II; Ortega, R.B. <u>Organometallics</u>, 1984, <u>3</u>, 1614.
29. Schilling, B.E.R.; Hoffmann, R.; Lichtenberger, D.L. <u>J. Am. Chem. Soc.</u> 1979, <u>101</u>, 585.
30. Kostic, N.M.; Fenske, R.F. <u>J. Am. Chem. Soc.</u> 1982, <u>104</u>, 3879-84.
31. Lichtenberger, D.L.; Fenske, R.F. <u>J. Am. Chem. Soc.</u> 1976, <u>98</u>, 50-63.
32. Lichtenberger, D.L.; Fenske, R.F. <u>Inorg. Chem.</u> 1976, <u>15</u>, 2015-22.
33. Lichtenberger, D.L.; Sellmann, D.; Fenske, R.F. <u>J. Organomet. Chem.</u> 1976, <u>117</u>, 253-64.
34. Kellogg, G.E. <u>Diss. Abstr. Int. B</u> 1986, <u>46</u>, 3838.
35. Calabro, D.C.; Lichtenberger, D.L. <u>J. Am. Chem. Soc.</u> 1981, <u>103</u>, 6846-6852.
36. Pang, L.S.K. <u>Diss. Abstr. Int. B</u> 1986, <u>46</u>, 3839-40.
37. Tatsumi, K.; Hoffmann, R.; Templeton, J.L. <u>Inorg. Chem.</u> 1982, <u>21</u>, 466-468.
38. Enemark, J.H.; Feltham, R.D. <u>Coord. Chem. Rev.</u> 1974, <u>13</u>, 339-406.

RECEIVED August 11, 1987

Author Index

Affiliation Index

Subject Index

A

Acetone, 15
Acoustic cavitation, 197
Acoustic ringing, 206
Adamantane, 167
Addition funnel pressure reactor, 201
Adjustable pressure relief valve, 200
Aerial oxidation, 64
Aerobic product transfer, 193
Aerosol pressure vessel, 198
Air-sensitive materials
  decomposition, 147
  HPLC analysis, 24
  recovering, 193
  synthesis and handling, 34
Alkyne electron density, 287
Alkyne ligand, 282
Alkyne π donor orbitals, 287
Alkyne π levels, 285
Ambient pressure flow cell, 238–244
Ammonia synthesis, 182
Anaerobic column chromatography, 17–18f
Anaerobic transfer, 144
Anionic polymerization, 182
Apparatus design philosophy, 117
Arc lamp
  collimated beam, 73
  vision damage, 70
Argon
  prepurified, 36
  purification, 83,84f
  vacuum line counterflow, 90
Atomic thermal parameters, 258
Azobenzene, 95

B

Ball and socket joint, 38,136
Ball valves, 157
Ballast tank, 107,108f
Band assignments, 274,285
Beam current, 164
Belt drive pumps, 118
Benzene, 15

Benzophenone, 85
Bercaw group, 79
Bimetallic transition metal catalysts, 182
Block copolymers, 182
Block length, 186
Bonding capabilities
  ligands, 273–274
  metal fragments, 284
Bonding character, 271
Bonding in organometallic molecules, 265–289
Butterfly valve, 121
Bypass designs, 119

C

Cannula techniques
  air-sensitive materials, 7
  components, 6
  double tip needles, 9
  flexibility, 7
  solution transfer, 6–23
  speed, 7
  stainless steel cannulae, 7
  teflon tubing, 9
Cannulation, low-temperature, 186
Carbon liner, 170
Carbon monoxide, 85
Carbonyl ligands, 99–100
Carbon–carbon bond, 287
Carbon–vapor reactor, 159
Catalytic system
  carbon monoxide and methanol, 233
  cyclohexanone, 238
Centrifuge-tube assembly, 255f
Ceramic fabrication, 179
Channelling, 38
Charge distribution, 271
Charge potential, 271
Chemical shift, 206,212
Chlorosilanes, 26
Chromatography
  anaerobic, 17
  cannula techniques, 17
  column evacuation, 147

Production by Cara Aldridge Young
Indexing by Colleen P. Stamm
Jacket design by Carla L. Clemens

Elements typeset by Hot Type Ltd., Washington, DC
Printed and bound by Maple Press, York, PA

Recent Books

Personal Computers for Scientists: A Byte at a Time
By Glenn I. Ouchi
276 pp; clothbound; ISBN 0–8412–1000–4

The ACS Style Guide: A Manual for Authors and Editors
Edited by Janet S. Dodd
264 pp; clothbound; ISBN 0–8412–0917–0

Silent Spring Revisited
Edited by Gino J. Marco, Robert M. Hollingworth, and William Durham
214 pp; clothbound; ISBN 0–8412–0980–4

Chemical Demonstrations: A Sourcebook for Teachers
By Lee R. Summerlin and James L. Ealy, Jr.
192 pp; spiral bound; ISBN 0–8412–0923–5

Phosphorus Chemistry in Everyday Living, Second Edition
By Arthur D. F. Toy and Edward N. Walsh
362 pp; clothbound; ISBN 0–8412–1002–0

Pharmacokinetics: Processes and Mathematics
By Peter G. Welling
ACS Monograph 185; 290 pp; ISBN 0–8412–0967–7

Detection and Data Analysis in Size Exclusion Chromatography
Edited by Theodore Provder
ACS Symposium Series 352; 307 pp; 0–8412–1429–8

Chemistry of High-Temperature Superconductors
Edited by David L. Nelson, M. Stanley Whittingham,
and Thomas F. George
ACS Symposium Series 351; 329 pp; 0–8412–1431–X

Reversible Polymeric Gels and Related Systems
Edited by Paul S. Russo
ACS Symposium Series 350; 292 pp; 0–8412–1415–8

Sources and Fates of Aquatic Pollutants
Edited by Ronald A. Hites and S. J. Eisenreich
Advances in Chemistry Series 216; 558 pp; ISBN 0–8412–0983–9

Nucleophilicity
Edited by J. Milton Harris and Samuel P. McManus
Advances in Chemistry Series 215; 494 pp; ISBN 0–8412–0952–9

For further information and a free catalog of ACS books, contact:
American Chemical Society
Distribution Office, Department 225
1155 16th Street, NW, Washington, DC 20036
Telephone 800-227-5558